Sustainable Textiles: Production, Processing, Manufacturing & Chemistry

Series Editor

Subramanian Senthilkannan Muthu, Chief Sustainability Officer
Green Story Inc.
Tung Chung, Hong Kong

This series aims to address all issues related to sustainability through the lifecycles of textiles from manufacturing to consumer behavior through sustainable disposal. Potential topics include but are not limited to: Environmental Footprints of Textile manufacturing; Environmental Life Cycle Assessment of Textile production; Environmental impact models of Textiles and Clothing Supply Chain; Clothing Supply Chain Sustainability; Carbon, energy and water footprints of textile products and in the clothing manufacturing chain; Functional life and reusability of textile products; Biodegradable textile products and the assessment of biodegradability; Waste management in textile industry; Pollution abatement in textile sector; Recycled textile materials and the evaluation of recycling; Consumer behavior in Sustainable Textiles; Eco-design in Clothing & Apparels; Sustainable polymers & fibers in Textiles; Sustainable waste water treatments in Textile manufacturing; Sustainable Textile Chemicals in Textile manufacturing. Innovative fibres, processes, methods and technologies for Sustainable textiles; Development of sustainable, eco-friendly textile products and processes; Environmental standards for textile industry; Modelling of environmental impacts of textile products; Green Chemistry, clean technology and their applications to textiles and clothing sector; Eco-production of Apparels, Energy and Water Efficient textiles. Sustainable Smart textiles & polymers, Sustainable Nano fibers and Textiles; Sustainable Innovations in Textile Chemistry & Manufacturing; Circular Economy, Advances in Sustainable Textiles Manufacturing; Sustainable Luxury & Craftsmanship; Zero Waste Textiles.

Subramanian Senthilkannan Muthu
Editor

Sustainable Apparel Retailing

Springer

Editor
Subramanian Senthilkannan Muthu
Green Story, Canada
Tung Chung, Hong Kong

ISSN 2662-7108 ISSN 2662-7116 (electronic)
Sustainable Textiles: Production, Processing, Manufacturing & Chemistry
ISBN 978-3-031-83874-3 ISBN 978-3-031-83875-0 (eBook)
https://doi.org/10.1007/978-3-031-83875-0

This Springer imprint is published by the registered company Springer Nature Switzerland AG
The registered company address is: Gewerbestrasse 11, 6330 Cham, Switzerland

This book is dedicated to:
The lotus feet of my beloved Lord
Pazhani Andavar
My beloved late father
My beloved mother
My beloved wife Karpagam and daughters
Anu and Karthika
My beloved brother
Last but not least
To everyone working in various industrial
spaces to make the world, "SUSTAINABLE"

Contents

Chapter 1
Sustainable Apparel Retailing

Subramanian Senthilkannan Muthu

Sustainable apparel retailing is emerging as a critical strategy for fashion retailers to address growing consumer demand for transparency and environmentally responsible products. As consumers become increasingly aware of the environmental and social impacts of the fashion industry, they are seeking out brands and retailers that align with their values around sustainability. Retailers are responding by implementing various sustainable practices throughout their operations and supply chains. This includes sourcing more sustainable and recycled materials, improving supply chain transparency, reducing waste and emissions, offering repair and recycling programs, and communicating their sustainability efforts to consumers through labeling and marketing. By adopting sustainable retailing strategies, fashion companies aim to build consumer trust, differentiate themselves in the market, mitigate risks, and contribute to a more circular and responsible fashion industry. While challenges remain, sustainable apparel retailing represents an important shift as the industry works to reduce its environmental footprint and meet evolving consumer expectations around ethical and eco-friendly fashion. This chapter will present an overview of the entire book.

The second chapter explains about a Conceptual Model of Sustainable and Responsible Fashion Consumption through Sustainability Labeling. This chapter introduces a conceptual model based on systems theory to demonstrate how sustainability labeling can be used by retailers to improve consumer knowledge and promote socially responsible fashion consumption. The model illustrates the interconnected relationships between sustainability labeling, consumer knowledge, purchase intention, and the four pillars of sustainability (economy, environment, equity, and esthetics).

S. S. Muthu (✉)
Green Story Inc., Canada, Tung Chung, Hong Kong

S. S. Muthu (ed.), *Sustainable Apparel Retailing*, Sustainable Textiles: Production, Processing, Manufacturing & Chemistry,
https://doi.org/10.1007/978-3-031-83875-0_1

This chapter begins with an overview of the current state of the retail industry and the growing consumer interest in sustainability. It then reviews relevant literature on socially responsible fashion consumption, barriers to such consumption, apparel labeling, and sustainability. The conceptual model is presented and evaluated using a case study of The Standard Clothing Company, a small apparel retailer that uses sustainability labeling. The results demonstrate how The Standard Clothing Company's sustainability labeling supports each pillar of sustainability while addressing consumers' desire for supply chain transparency and sustainable consumption options. This chapter concludes by discussing the implications of the model for retailers and suggesting areas for future research, including empirically testing the relationships proposed in the model.

The third chapter explains about Designing Sustainable fashion for circular behavior.The fashion industry is liable for producing huge amount of waste every year causing the problem of landfill wastes and overconsumption. The current business model of the fashion industry is unsustainable as it supports overconsumption and generates unrestricted landfill waste. Previous literature pointed out that consumers generate half of the impact of the fashion industry. Fast fashion is one of the main reasons behind the problem of overconsumption. The main aim of this study is to explore the sustainable design strategies for development of garments in terms of life extension, redesign, and in changing the consumption behavior of consumers. Sustainable design aims to find new ways of planning waste and producing ethical products. Researchers used terms such as reduce, sustain, halt, avoid, and limit and refer to finding out ways that are less likely to attract criticism than the conventional ways. These strategies emphasize reducing the defects of the current system rather than changing the whole system. Sustainable practices can be implemented both in the pre-consumer and post-consumer stage for waste minimization. Common waste management strategies are reduce, reuse, and recycle and their main purpose is to prolong the lifecycle of a product. This study also explore the concept of circular fashion along with design of circularity, design for disassembly, and zero waste fashion along with waste management strategies. This study will be an attempt to explore all the strategies concerning their impact on circular behavior.

The fourth chapter is focused on the Present Challenges and Future Opportunities of Sustainable Retailing in the Apparel Industry. Sustainable apparel retailing plays a pivotal role in addressing environmental and social concerns, making it imperative to understand the challenges, leverage opportunities, and anticipate future trends in this dynamic sector. This chapter explores a comprehensive analysis of the complex landscape of sustainable apparel retail, elucidating the array of challenges, opportunities, regulatory frameworks, case studies, and future trends shaping this dynamic sector. Through an in-depth examination, it investigates the critical challenges confronting retailers, including sourcing sustainable materials, ensuring transparency and traceability in the supply chain, navigating cost implications and reconciling sustainability objectives with prevailing fashion trends and design paradigms. Conversely, it delineates the promising avenues available to retailers, encompassing innovative strategies related to materials sourcing and manufacturing processes, strategic collaborations with sustainable fashion stakeholders, nuanced marketing and communication approaches, and introducing sustainable packaging and logistics solutions. Moreover, this chapter thoroughly examines the regulatory and certification frameworks governing sustainable apparel retailing, providing insights into prominent sustainability certifications such as Fair Trade, GOTS, and B Corp. It further delves into considerations regarding compliance with environmental and labor standards alongside pertinent legal considerations. Additionally, through a meticulous analysis of case studies and best practices, the paper illuminates' successful initiatives in sustainable apparel retailing, drawing from exemplary endeavors by leading sustainable fashion entities and offering tailored strategies for effectively adopting sustainable practices by small and medium-sized retailers. Lastly, this chapter casts a forward-looking perspective, elucidating emerging technologies and innovations, evolving consumer preferences, market dynamics, and avenues for growth and expansion within the sustainable apparel retail sector. This study offers a robust framework for comprehensively understanding and navigating the challenges, leveraging the opportunities, and strategically positioning oneself amidst the evolving landscape of sustainable apparel retail.

The fifth chapter explains about the Circular Packaging Models in Apparel Retailing. Packaging is an important part of apparel retailing because it protects apparel products during transit. In recent times, packaging materials are expected to be visually appealing and represent the brand's identity rather than protecting the product. Recent research found that 8.3 billion tons of plastic have been produced worldwide since the 1950s of which 60% of it ended up being discarded (Rhodes 2019). Packaging materials are mainly made of polyethylene and polypropylene which are technically plastics that have an enormous contribution in producing poly waste that harms the environment. Forming a sustainable solution in packaging, the term 7Rs was coined. The 7Rs provide a thorough framework for a business to move forward with a more circular packaging model. Many businesses have taken up the initiative of following the 7Rs strategy to improve their overall packaging experience and sustainability. For example, re-pack is an initiative started by many businesses where consumers deliver the packaging material back to the organization after receiving the product from an online order. Based on this concept, lots of tools have been developed that help companies develop sustainable packaging products with the help of proper channels. One such example is PIQET, which is approved by the "Nation Packaging Corporation" and helps businesses assess their package's environmental footprint. These tools will also help consumers make informed choices about the purchases they make as they show the impact of the packaging material on the environment. In this chapter, the different types of apparel packaging, materials used, environmental impact, waste generated, sustainable models for packaging like 7R, and sustainable packaging alternatives are discussed.

The sixth chapter analyses different strategies for adopting sustainable apparel retail practices. Consumers are one of the most influential external stakeholders of any apparel retail firm. Being considered with such importance, most of them are now getting to understand the significance of the products they use regarding sustainability factors. These factors uplift not only the suppliers but also the retailers to inculcate a sustainable culture in apparel retailing. As the apparel sector is influenced by a large supply chain throughout its entire processing, the focus on green retail practices gives scope to improve the supply chain management to a greater extent. Effective communication between the retailers and suppliers is imperative to elucidate a good medium for the consumers to interpret sustainability effortlessly. This paper studies how consumers' sustainability principles are driven toward the retail transition from brick-and-mortar to e-commerce. The relationship between the retailer and suppliers should be well managed to establish sustainable practices in the apparel industry. Even when the users understand the efforts of the retailers toward sustainability goals, the retailers are not motivated due to users' behavior as they do not give more preference to sustainable products and it should be changed to ensure effective sustainable retailing practices. The key strategies to maintain sustainable retailing are discussed in the view of achieving long-term environmental and social objectives while maintaining profitability. This chapter discusses in detail about the various strategies applicable to ensuring sustainability in apparel retailing and the sustainability outcomes of the proposed activities are also highlighted.

The seventh chapter explains about the Sustainable Reverse Logistics Practices in Apparel Retailing. Sustainability in fashion encompasses many factors, including cutting CO_2 emissions, addressing overproduction, reducing pollution and waste, supporting biodiversity, and ensuring that garment workers are paid a fair wage and have safe working conditions. An effective supply chain is essential to address the sustainability issues in the fashion industry. There is also a need for handling the reverse supply chain and that is the moving of products from customer to manufacturer or disposal for many reasons. Reverse logistics is the process that manages this. It needs to be sustainable and efficiently used in the fashion supply chain to effectively handle the returned goods or materials without decreasing their value and to increase the scope for further usage. It involves a return policy between retailers and manufacturers, where retailers receive monetary support for returning unsold products and subsidizing some loss at the end of the selling season. This policy can improve supply chain coordination and benefit both retailers and suppliers. It also occurs between consumers and retailers in five dimensions, which are explained in this chapter. The return policies can attract more consumers and create more loyal customers, reduce consumer risks, increase demand, lower transaction costs, and increase customer value. However, research from a profitability perspective suggests that return policies may not maximize profitability. Reusing and recycling practices in the global fashion industry vary across different cases. The development of information technology contributes to the sustainability of fashion e-retailing and effective reverse logistics. In this chapter, the concept of reverse logistics, its procedures, and processes, challenges, growth areas, and sustainable practices to be incorporated into reverse logistics, various brands focusing on reverse logistics, and emerging opportunities in reverse logistics to ensure sustainability in apparel retailing are discussed.

The eighth chapter explains about facilitating the sustainable conversations: toward a human-like chatbot for the fashion industry. This chapter delves into the development of a novel conceptual framework for AI-powered chatbots aimed at promoting sustainability in the fashion industry. As the use of chatbots becomes increasingly common in e-commerce, particularly in enhancing customer engagement and personalization, their potential for driving sustainable consumer behaviors remains underutilized. This chapter introduces a framework that integrates multichannel strategies and human-like features into chatbot design to enhance value co-creation and raise awareness of sustainable practices among consumers. This innovative approach addresses a critical gap in the fashion industry, where sustainable consumption and circular fashion need more focused promotion. The framework suggests that a delicate balance between human-like and robotic characteristics in chatbots can effectively influence consumer attitudes toward sustainability. By providing practical insights for fashion marketing managers, the framework aims to transform how brands communicate sustainability, ultimately leading to a more informed and conscientious consumer base. This chapter also outlines future research directions, emphasizing the potential impact of such chatbots in fostering a culture of sustainability within the fashion industry.

I take this opportunity to thank all the contributors for their earnest efforts to bring out this book successfully. I am sure readers of this book will find it very useful.

Chapter 2
From Retailer to Consumer and Back Again: A Conceptual Model of Sustainable and Responsible Fashion Consumption Through Sustainability Labeling

Annie Williams, Nancy Hodges, and Kittichai (Tu) Watchravesringkan

Recently, the retail industry is said to be experiencing an "apocalypse," accelerated by the coronavirus, in that brick-and-mortar stores and shopping malls are rapidly closing down as consumers prefer to make purchases online (Rankin et al., 2020). However, some contend that the retail industry is simply experiencing a reinvention, as it reorganizes itself to better appeal to consumers (Wang, 2020). Prior research suggests that successful retail reinvention includes a focus on consumer's need for experience, convenience, discovery of new products and value, while adopting more omnichannel strategies and advanced technologies to garner consumer support (Fiedler et al., 2020; Rankin et al., 2020; Wang, 2020). However, one overlooked solution to retail reinvention is to address consumer's desire for access to sustainability information and their interest in consuming sustainable products. Indeed, Sides and Skelly (2021) contend that to garner consumer loyalty during this new phase of retail, companies should integrate transparency into the entire supply chain network. According to one survey, 46% of respondents reported trying new brands and/or retailers that exhibited socially responsible practices in response to the pandemic (Fiedler et al., 2020). Similarly, a 2017 pre-pandemic consumer survey reports that 87% of Americans want to buy a product with a social and/or environmental benefit, and 89% of the population is willing to switch brands, holding price and quality constant, to do so (Cone Communications, 2017). Post-pandemic,

A. Williams (✉)
Kenan-Flagler Business School, University of North Carolina at Chapel Hill, Chapel Hill, NC, USA
e-mail: annie_williams@kenan-flagler.unc.edu

N. Hodges · K. (Tu). Watchravesringkan
Department of Consumer, Apparel and Retail Studies, University of North Carolina at Greensboro, Greensboro, NC, USA
e-mail: njnelson@uncg.edu; k_watchr@uncg.edu

S. S. Muthu (ed.), *Sustainable Apparel Retailing*, Sustainable Textiles: Production, Processing, Manufacturing & Chemistry,
https://doi.org/10.1007/978-3-031-83875-0_2

previous research notes that nearly half of United States' consumers are concerned about sustainability (Alldredge & Teichner, 2023) and are increasingly supporting brands that offer sustainable, affordable options (Das et al., 2023).

Promoting sustainability not only facilitates consumers' ability to engage in sustainable consumption, but also benefits the retailer, as past research demonstrates that retailers engaged in sustainability benefit from a competitive advantage including greater consumer store evaluations (Hofenk et al., 2019) as well as increased satisfaction (Gil-Saura et al., 2023), store equity, customer loyalty, positive word-of-mouth communication, and consumer intention to shop with the retailer (Lee, 2021; Sánchez-González et al., 2020). Moreover, retailers serve a critical position with regards to sustainability, as their position between suppliers and consumers offers an opportunity to facilitate a circular economy, as well as promote sustainable practices from both suppliers and consumers (Vadakkepatt et al., 2021). As such, retailers are capable of promoting sustainable consumption and often appeal to consumers' needs for sustainability (Lehner, 2014).

Such efforts are needed to improve upon the unsustainable nature of the global fashion industry given that it is estimated to be responsible for nearly 4% of the world's greenhouse gas emissions, a figure that is equivalent to Germany, France, and the United Kingdom's combined annual carbon emissions (Berg et al., 2020). Without any changes to current consumption or production practices, it is estimated that by 2030, the climate impact of these industries will increase by 49%, which is equivalent to the current climate impact of the entire USA (Chrobot et al., 2018). As such, the industry's current production practices pose a threat to the world's collective efforts to cap the increase in global temperature to 1.5°C above pre-industrial levels, which is the maximum increase in global temperature allowed before risking irreversible climate change (Berg et al., 2020).

Impact on climate change is just one example of the unsustainable nature of the apparel industry. Others include subjecting employees to poor labor conditions, such as low wages, long hours, and dangerous working conditions (Ross & Morgan, 2015). Moreover, apparel production has resulted in increased apparel and textile waste in landfills (Ross & Morgan, 2015), as in 2018 alone, an estimated 11 million tons, out of the estimated 17 million tons of apparel and textiles produced in the USA that year, were landfilled (United States Environmental Protection Agency, n.d.). Furthermore, cotton demand for apparel production is depleting natural water resources (Gupta & Gentry, 2018), while pesticides and chemicals used in cotton cultivation and fiber treatment are contaminating freshwater streams (Ross & Morgan, 2015). Additionally, frequent shipping of apparel is contributing to heightened carbon dioxide emissions (Turker & Altuntas, 2014). Similarly, increased deforestation, driven by the expansion of land for textile fiber cultivation, is impeding the ability to sequester carbon dioxide emissions, as forests serve as natural carbon sinks (Intergovernmental Panel on Climate Change, 2014).

Despite a reported consumer desire to consume apparel sustainably, an intention-behavior gap exists, wherein consumers fail to purchase products with such benefits (James & Montgomery, 2017a). This gap is due to such barriers as a lack of consumer knowledge concerning the ethical issues that surround fast fashion apparel

(Gupta & Gentry, 2018), lack of awareness about where to obtain accurate information regarding the sustainability of products (James & Montgomery, 2017b), and lack of ability to judge how sustainable products really are (McNeill & Moore, 2015). Thus, consumers report wanting more transparency regarding fashion production to overcome these barriers (Sustainable Apparel Coalition, 2019). Specifically, consumers want to know more about how and where the clothing was sourced, the impact of the fibers that comprise their garments, as well as the working conditions of apparel factories (Sustainable Apparel Coalition, 2019).

Given the growing concern regarding the sustainability of apparel among consumers (Sustainable Apparel Coalition, 2019), and a desire to consume products with a social benefit (Alldredge & Teichner, 2023; Cone Communications, 2017; Das et al., 2023; Fiedler et al., 2020) there lies an opportunity for fashion retailers to reinvent themselves by appealing to consumer's interests in supply chain transparency and enabling consumers to make informed purchasing decisions by implementing sustainability labeling on apparel. Doing so would allow fashion retailers to be the link between apparel producers and consumers, delivering the apparel production transparency that consumers' desire. Such an initiative would reinvent the fashion retail industry as we know it, given that no universal, sustainability-related apparel labeling currently exists (Sustainable Apparel Coalition, 2019). While some may argue that such labeling will only benefit those retailers with existing sustainable supply chains, the Sustainable Apparel Coalition (2019) reports that many retailers and brands can garner consumer trust by being honest about how their products are produced and how negative issues are addressed.

Although previous research has focused on the implementation of ecological and social labels on apparel (Baker, 2002; Hilowitz, 1997; Koszewska, 2011) and their effect on attitude toward the brand and purchase intention of the product (D'Souza et al., 2006; Dickson, 2001; Hustvedt et al., 2008; Hyllegard et al., 2012), there is little research that demonstrates how sustainability labeling appeals to consumer's interest in accessing apparel sustainability information enough to influence more sustainable and responsible consumption and therefore to support the pillars of sustainability. It is necessary that such labeling actually supports sustainability so that retailers can avoid "corporate hypocrisy" (James & Montgomery 2017a, p. 73) or "...when companies make false or exaggerated claims about how environmentally friendly their products are" (Solomon, 2017, p. 54), also known as greenwashing.

The purpose of this chapter is to introduce a conceptual model based on systems theory and to evaluate the model using a case study approach. The model offers fashion retailers an opportunity to reinvent their retail operations by appealing to the consumer's desire for sustainability information related to apparel production (Sustainable Apparel Coalition, 2019), and interest in consuming sustainable products (Cone Communications, 2017). Moreover, conceptually, systems theory allows for an investigation of the interdependent relationships among sustainability labels, consumer knowledge, socially responsible fashion consumption and sustainability. Such an investigation offers insight regarding the leverage points necessary to positively influence socially responsible fashion consumption, thereby meeting the consumer's sustainable apparel consumption needs.

This chapter proceeds as follows: first the relevant literature pertaining to socially responsible fashion consumption, barriers to such consumption, apparel labeling and its effects on consumer behavior, as well as sustainability, and systems theory is presented. The conceptual model is then introduced. This section is followed by the application of one retailer's sustainability labeling efforts as a case study in order to assess the conceptual model. Lastly, conclusions, limitations, and areas of future research are presented.

2.1 Literature Review

2.1.1 Socially Responsible Fashion Consumption

A person who engages in socially responsible consumer behavior is "... one who takes into account the public consequences of his or her private consumption or who attempts to use his or her purchasing power to bring about social change" (Webster Jr., 1975, p. 188). While socially responsible consumer behavior has been defined, there is currently no agreed upon definition for socially responsible consumer behavior in the context of fashion consumption (Kozlowski et al., 2018). There is, however, consensus that sustainable business in general be built on the foundation of sustainable development as outlined in the 1987 Brundtland report (Kozlowski et al., 2018; Stanszus & Iran, 2015).

The Brundtland report defines sustainable development as "... development that meets the needs of the present without compromising the ability of future generations to meet their own needs" (United Nations, 1987, para. 55) and emphasizes the inclusion of environmental protection, economic growth, and social equity for sustainable development (United Nations, 1987). As such, Kozlowski et al. (2018) define sustainable fashion as "the profitable design, production, distribution, and end-of-life reuse, recycling, or disposal of fashion that supports circular systems, minimizes negative and maximizes positive impacts on both society and the natural environment" (p. 195). Combining Webster's (1975) and Kozlowski et al.'s (2018) definitions, considering their focus on improving society, a socially responsible consumer in the context of fashion consumption is a consumer that engages in the consumption of sustainable fashion or clothing as defined by Kozlowski et al. (2018). Therefore, socially responsible fashion consumption, or SRFC, is defined as the acquisition, use, and/or disposal of fashion items that supports circular systems and minimizes negative and/or maximizes positive impacts on society and the natural environment.

2.1.2 Barriers to Socially Responsible Fashion Consumption

Despite consumer interest in products with a social and/or environmental benefit (Alldredge & Teichner, 2023; Cone Communications, 2017; Das et al., 2023; Fiedler et al., 2020), there are noted barriers to actually engaging in SRFC that result in the aforementioned intention-behavior gap. One such barrier is consumer knowledge, as some consumers do not know about the ethical issues that surround their apparel purchases (James & Montgomery, 2017a). Of those that do, many do not know where to obtain accurate information regarding the sustainability of products (James & Montgomery, 2017a), how sustainable the products are that they purchase (McNeill & Moore, 2015; Williams & Hodges, 2022b), where to purchase sustainably sourced products (James & Montgomery, 2017b; Williams & Hodges, 2022a), or the social responsibility record of the companies they patronize (Mohr et al., 2001; Williams & Hodges, 2022a). Others are not familiar with the complex nature of the apparel supply chain (James & Montgomery, 2017a) and the social and environmental effects of clothing production and consumption (Connell, 2010; Harris et al., 2016; Hill & Lee, 2012; James & Montgomery, 2017a). Moreover, many report lacking knowledge regarding how and where to properly dispose of or recycle clothing (Goworek et al., 2012; Morgan & Birtwistle, 2009; Rathinamoorthy, 2020). Past research also demonstrates that consumers do not have enough information concerning the social responsibility performance of the companies they patronize (Mohr et al., 2001). Thus, while consumers may have positive intentions to engage in SRFC, the aforementioned studies demonstrate that many lack the knowledge to be able to do so.

2.1.3 Apparel Labeling

To overcome the aforementioned barriers and to encourage SRFC, consumers and producers could utilize apparel labeling. Hilowitz (1997) defines a label as "information that the manufacturer or marketer of a product provides to the consumer at the point of sale" (p. 216). Apparel producers use labels to differentiate their products from competitors by communicating information related to the quality and benefits of their product (Baker, 2002). Moreover, labels are used by apparel manufacturers to communicate information about a product to a consumer, thereby enabling them to make informed purchasing decisions (Hyllegard et al., 2012).

In the USA, all apparel items must be labeled with "the fiber content, the country of origin, the manufacturer or dealer identity, and the care instructions" (OTEXA, n.d., para. 1). Such information should be affixed to the product until delivered to the consumer and should be clear, visible, and easy for consumers to access (Federal Trade Commission, n.d., para. 29). This required information can appear on one or more labels, however, all parts of the fiber content information must appear close to each other, and the type and lettering must be of equal size and conspicuousness (Federal Trade Commission, n.d., para. 32–33).

Beyond the requirements of the United States government, some brands may choose to add additional labels, such as environmental, ecological, or social labels to differentiate their products from competitors, as doing so communicates environmental or social responsibility (D'Souza, 2000; Koszewska, 2011). Environmental labels are known to make consumers aware of environmental issues, improve the image of a company, and encourage companies to account for the effects of their production (Gallastegui, 2002).

Commonly, brands utilize the International Organization for Standardization guidelines to communicate the environmental friendliness of their products. The International Organization for Standardization has divided environmental labels into three types: Type I, Type II, and Type III (Koszewska, 2015). Type I, II, and III classifications are only applied to environmental labels; as such, classifications are based on life-cycle analysis (Koszewska, 2015) or the environmental impacts of a product over its life cycle, from raw material to disposition (International Organization for Standardization, 2006).

Type I labels are multi-criteria based, third party certified and based on a simplified life-cycle analysis (Koszewska, 2015). Examples of such labels include the Blue Angel Label, the European Union Eco-Label (Gallastegui, 2002), and Nordic Swan. These labels communicate that a product is environmentally friendly throughout its entire life cycle. *Type II* labels are claims made by manufacturers, importers, or distributors that refer to a single, specific product attribute (Gallastegui, 2002; Koszewska, 2015). Examples of ISO *Type II* labels are labels that attest that products are not tested on animals and CFC-free (Koszewska, 2011). *Type III* labels use pre-set indices to assess a product's entire life cycle and are certified by a third party (Koszewska, 2015). Examples include nutrition labeling on food and environmental product declaration labels (Koszewska, 2011). While the International Organization for Standardization publishes the guidelines firms should adhere to in order to use the *Type I, II*, or *III* labels, the International Organization for Standardization does not provide certifications. Instead, such certifications are provided by a third-party agency.

While ecological labels focus on the environment, social labels "...inform consumers about the social conditions of production in order to assure them [consumers] that the item or service they are purchasing is produced under equitable working conditions" (Hilowitz, 1997, p. 216). Most social labels communicate the living and working conditions of laborers in developing countries (Hilowitz, 1997). As such, social labels often include equity-related aspects of an item's production, including fair-trade conditions and worker rights (Koszewska, 2011). There is not a universal certification system for social responsibility; however, some brands voluntarily use the following labels: Clean Clothes Campaign, Rugmark (Hilowitz, 1997; Koszewska, 2011), Care and Fair, Fair Wear Foundation, and the Oeko-Tex Standard 100 labels. Each of these labels maintains different standards, and thereby communicates different aspects of social responsibility. For example, the Clean Clothes Campaign label signifies that a brand is a part of a global network committed to fair working conditions for apparel workers (Clean Clothes Campaign, n.d.). The Rugmark and Care and Fair labels certify that products are made without illegal child labor (Care and Fair, n.d.; Rugmark, n.d.). The Fair Wear Foundation label certifies that a product is made by employees who have a right to collective bargaining, are not discriminated against, are paid a living wage, and have reasonable work hours and safe working conditions (Fair Wear, n.d.). The Oeko-Tex Standard 100 label guarantees that all components of a product (e.g., threads, buttons, etc.) are harmless to human health (Oeko-Tek Standard 100, n.d.).

While the International Organization for Standardization developed *Type I, II,* and *III* labels for environmental labels, they are based on garment life-cycle analyses (Koszewska, 2015). Because it is difficult to analyze the life cycle of a garment in terms of its social impact (Koszewska, 2015), *Type I, II,* and *III* certifications are not used for social labels. However, the International Organization for Standardization does use the 26,000 classification to offer organizations standards for social responsibility (International Organization for Standardization, 2010). That is, those organizations that meet the International Organization for Standardization's guidelines for social responsibility can appeal to a third-party certification agency for the use of the ISO 26000 label (International Organization for Standardization, 2010). In addition to ecological and social labels, apparel producers may also use sustainability labels on products. Such labels are those that communicate both a product's ecological and social impact (Koszewska, 2015). Examples include the Global Organic Textile Standard and Cradle to Cradle (Koszewska, 2015), Fair-Trade, Oeko-Tex Made in Green, and BLUESIGN labels. Similar to social labels, each of these labels maintains different requirements for its use. The BLUESIGN label signifies that a product is manufactured with the lowest possible impact on people and the environment (Bluesign, n.d.). The Global Organic Textile Standard label illustrates that a product is made in accordance with environmental and social standards, such as the use of organic fibers and no use of child labor (Global Organic Textile Standard, n.d.). The Cradle-to-Cradle label indicates that a product is made with materials healthy for humans and the environment, as well as uses methods that promote product circularity, clean air and climate protection, water and soil stewardship and social fairness (Cradle to Cradle, n.d.). The Fair-Trade label indicates that a product was made in safe working conditions using methods that support environmental protection and sustainable livelihoods (Fair Trade, n.d.). Lastly, the Oeko-Tex Made in Green label communicates that a product is made without harmful substances and in a socially responsible way (Oeko-Tex Made in Green, n.d.). Table 2.1 provides a review of commonly used environmental, social, and sustainable apparel labels, as well as relevant apparel labeling associations.

Despite a consumer interest in accessing more information concerning the production of the apparel they purchase (Amed et al., 2019; Ditty, 2020; Goworek et al., 2012; Sustainable Apparel Coalition, 2019; Williams & Hodges, 2022b; Vehmas et al., 2018; Yudina, 2017), the apparel industry has yet to implement a universal label to communicate such details (Radhakrishnan, 2014). Nevertheless, according to the literature, consumers prioritize the following apparel production information: material quality (Sustainable Apparel Coalition, 2019), labor conditions (Modi & Zhao, 2021; Williams & Hodges, 2022b) and the environmental impact of apparel production (Amed et al., 2019; Ditty, 2020). Studies have found that consumers prefer such information to be communicated in an explicit, succinct,

Table 2.1 Examples of Apparel Labels and Labeling Associations

Name/Source	Type	Description	Logo
Blue angel www.blauer-engel.de/en	Ecological-ISO type I	Eco-label of Germany that certifies the environmental friendliness of a breadth of products across their entire lifecycle. More than 20,000 products are certified by blue angel including textiles, clothing, and shoes.	
BLUESIGN www.bluesign.com/	Sustainable	The BLUESIGN logo ensures that textile consumer products are manufactured with the lowest possible impact on people and the environment.	
Care and fair www.care-fair.org	Social	The care and fair initiative works against the use of illegal child labor in the carpet industry.	
Clean clothes Campaign https://cleanclothes.org/	Social	The clean clothes Campaign is a network of more than 235 organizations located in over 45 countries that focuses on the protection of laborers in the garment and sportswear industries.	
Cradle to cradle www.c2ccertified.org/	Sustainable	The cradle to cradle program promotes the use of safe, circular, and responsibly made products. Certification addresses five dimensions: That materials used are healthy for humans and the environment, product circularity, clean air and climate protection, water and soil stewardship and social fairness. Products include apparel, shoes, and textiles.	
EU eco-label www.ecolabel.eu	Ecological-ISO type I	The EU eco-label certifies the environmental friendliness of a diverse range of products and services, attesting they are environmentally friendly across their entire lifecycle. Products include clothing, textiles, and footwear.	
Fair-Trade certification www.fairtradecertified.org	Sustainable	The fair-Trade certification addresses social, environmental, and economic standards, ensuring that the production of fair-Trade certified products is in accordance with safe working conditions, environmental protection, and sustainable livelihoods. Products include food, clothing, flowers, shoes, home goods and beauty products.	

(continued)

Table 2.1 (continued)

Name/Source	Type	Description	Logo
Fair Wear foundation www.fairwear.org/	Social	The fair Wear foundation is an independent organization of more than 147 brands focused on improving the working conditions of garment laborers.	FAIR WEAR
Global organic textile standard (GOTS) https://global-standard.org/	Sustainable	GOTS ensures that the production of textiles is in accordance with environmental and social standards.	GLOBAL ORGANIC TEXTILE STANDARD · GOTS ·
Nordic swan www.nordic-ecolabel.org/	Ecological-ISO type I	The Nordic swan is the official eco-label of the Nordic countries. Testing products across their entire lifecycle, the Nordic swan logo attests that certified products are environmentally friendly.	
Oeko-Tex standard 100 www.oeko-tex.com/	Social	The Oeko-Tex standard 100 attests that every component of a textile article, including threads, buttons, and other accessories, are free from substances harmful to humans.	OEKO-TEX® CONFIDENCE IN TEXTILES STANDARD 100
Oeko-Tex made in green www.oeko-tex.com/	Sustainable	The Oeko-Tex made in green logo attests that textiles are made without harmful substances and in socially responsible ways.	OEKO-TEX® INSPIRING CONFIDENCE MADE IN GREEN
Rugmark https://rugmarkindia.de/	Social	The Rugmark certification ensures that carpets have been manufactured and exported by companies that do not use illegal child labor.	RUGMARK

and easily comprehensible manner (Han et al., 2017b; Hyllegard et al., 2012; Ma et al., 2017; McNeill & Moore, 2015; Williams & Hodges, 2022b; Yan et al., 2012). Effective communication methods include narratives (e.g., stories about who and/or how the apparel item was made), logos, hangtags (Hyllegard et al., 2012; Lee et al., 2020; Li & Leonas, 2021; Sustainable Apparel Coalition, 2019), and such digital platforms as social media, websites, product labels, or QR codes (Sustainable Apparel Coalition, 2019; Williams & Hodges, 2022b). Findings of past research suggest that producers should use multiple communication channels and media to engage consumers (Han et al., 2017b; Vehmas et al., 2018).

2.1.4 Labeling and Consumer Behavior

The literature regarding whether consumers access and utilize the information provided on apparel labels is mixed. Hyllegard et al. (2012) found that 60% of American respondents read hangtags on a frequent or very frequent basis. When it comes to utilizing the care label, three quarters of American women consult a garment's label for care information, factoring this into the cost of a garment prior to purchasing it (Label Consciousness, 2007). In a Canadian study, half of participants reported using the information on care labels almost always or always when choosing apparel (Feltham & Martin, 2006) and in a South African study, 52% of participants always read labels, indicating that their purchase decisions were affected by the information displayed on the label (Van Der Merwe et al., 2014). However, in a U.K. based study, Iwanow et al. (2005) found that only 11% of consumers frequently looked at the label of branded apparel prior to purchasing it, and in a Norwegian based study (Laitala & Klepp, 2013), not all participants read the care labels, with many relying on their previous experience with similar garments to care for clothing.

Previous studies have demonstrated that environmental, social, and sustainability labels enable consumers to make more environmentally and socially responsible purchase decisions (D'Souza et al., 2006; Hyllegard et al., 2012; Ma et al., 2017). Indeed, prior research suggests that such labels positively affect consumer knowledge, and subsequently their purchase intention of sustainably-labeled apparel. For example, the more knowledge consumers have, the more likely they are to overcome barriers associated with consuming environmentally friendly textiles and apparel (Kang et al., 2013). Similarly, prior research suggests that consumers with more knowledge regarding the ethical issues surrounding apparel production are more likely to purchase environmentally and socially responsible apparel (Blazquez et al., 2020; Cowan & Kinley, 2014; Diddi & Niehm, 2016; Goworek et al., 2012; Ko & Jin, 2017; Kozar & Connell, 2013; Okur & Saricam, 2018; Zheng & Chi, 2015). Likewise, retailers who offer consumers the opportunity to engage in SRFC positively influence consumers' SRFC knowledge, which, in turn, positively influences their intention to engage in SRFC (Han et al., 2017a). Moreover, some research has demonstrated that consumers with more apparel production knowledge have more favorable attitudes toward organic cotton and are more willing to buy organic cotton clothing at higher price points than those with less knowledge (Oh & Abraham 2015). Despite these findings, there are also past studies suggesting that knowledge concerning the production practices of apparel does not always lead to sustainable consumption behavior (Brosdahl & Carpenter, 2010; Connell & Kozar, 2012; Iwanow et al., 2005; Kim & Damhorst, 1998; Kozar & Connell, 2010).

Nevertheless, prior research suggests that to facilitate consumer's access to knowledge related to sustainability and apparel, brands should focus their efforts at the retail level, between the time that consumers are deciding among competing products and the time they are purchasing them (James & Montgomery, 2017b). In between these two stages, a window of opportunity emerges whereby retailers can influence consumer's purchasing decisions (James & Montgomery, 2017b). Typically, retailers use this window of opportunity to upsell an item or to influence consumers to purchase additional goods (James & Montgomery, 2017b). However, retailers could use this window of opportunity to implement messaging to inform consumers regarding the ethical and sustainable issues surrounding their fashion purchase to promote SRFC (James & Montgomery, 2017b). Past research indicates such initiatives have been successful in influencing consumers to purchase more sustainable food items (Jäger & Weber, 2020).

Beyond the effect of apparel labeling on consumer's knowledge of SRFC, previous studies also indicate a positive effect of apparel labeling on consumer attitudes toward the label itself, which, in turn, positively affects consumer attitudes toward the purchase of sustainably-labeled apparel. Based on the Technology Acceptance Model, Ma et al. (2017) found that sustainability apparel labels that are perceived to be easy to use (i.e., learning the label is easy, the label is easy to make purchase decisions, and it would be easy to become an expert at using the label) and useful (i.e., the label gives access to useful information, the labels will make purchasing easier, and purchase decisions would be difficult without the label) positively influence attitudes toward the use of the label, which, in turn, positively influences consumer purchase intention of sustainably-labeled apparel. Similarly, Hyllegard et al. (2012) determined that consumers have more favorable attitudes toward hangtags with prosocial, explicit claims than hangtags without such claims. In Hyllegard et al.'s (2012) experiment, explicit claims were communicated, including that the garment was made without exploitive practices (e.g., sweatshops, unfair wages or child labor), instead of simply stating that the garment was made with fair labor practices. Consumers' positive attitudes toward these prosocial, explicit hangtags, in turn, positively influenced their purchase intention of socially-labeled apparel (Hyllegard et al., 2012). Hyllegard et al. (2014) reached similar conclusions, finding that college students preferred t-shirts with hangtags containing prosocial claims vs. hangtags without such claims. Prosocial hangtags, in turn, influenced their attitudes and purchase intention of socially-labeled apparel (Hyllegard et al., 2014).

A direct, positive relationship between social labeling and purchase intention of sustainably-labeled apparel has also been found to exist. For example, Dickson (2001) found that participants' purchase intention of sustainable apparel was influenced by the inclusion of a no-sweat label, which indicates that the apparel was made in fair working conditions. Similarly, Hustvedt et al. (2008) suggested that, in some cases, consumers are motivated to purchase apparel products that are labeled for animal welfare. Positive consumer purchase intention also resulted from apparel items labeled as containing organic, non-genetically modified or locally-produced fibers (Hustvedt & Bernard, 2008). In addition, previous studies suggest that apparel products labeled as environmentally friendly, ethically sourced, produced with 100% cotton or organic resulted in positive purchase intentions (Byrd & Su, 2021). Some consumers are willing to pay more for apparel products labeled as organic, natural, eco-friendly, sustainable (Bernard et al., 2013) or that the apparel product was made in accordance with international labor laws by workers that are a part of an independent trade union (Hustvedt & Bernard, 2010).

Past studies also suggest that brands benefit from the use of social and sustainable labeling. For instance, Yan et al. (2010) found that participants developed more favorable attitudes toward brands that used labeling to depict the fair treatment of workers, as opposed to brands that promoted sexual or dual (sexual and fair treatment of workers) appeals. Similarly, Hyllegard et al. (2012) and Williams et al. (2024) found that the use of labeling resulted in positive attitudes toward the brand. Likewise, brands that utilize an apparel sustainability index label to communicate their sustainable apparel offerings benefit from more positive brand evaluations including dimensions of brand equity and brand resonance (Williams et al., 2024).

2.1.5 *Sustainability*

Many definitions for sustainability abound, but a recurring theme in most definitions is that they are rooted in both the Iroquois' Great Law and the United Nations' (1987) Brundtland Report's concept that sustainable actions are those that allow for the current population to meet their own needs without compromising the ability of future generations to do the same (Niesenbaum, 2020). Niesenbaum (2020) extends the definition of sustainability by contending that sustainability is best achieved by using systems thinking, by back-casting principles to ensure the responsible use of resources, and to secure equitable economic opportunity and an improved standard of living for all people.

The Brundtland Report (United Nations, 1987) maintains that there are three pillars of sustainability, or sustainable development: "economic growth, environmental protection, and social equality" (Niesenbaum, 2020, 27). A sustainable economy, or ecological economy, is one that promotes circular systems, functions within its ecological limits, encourages renewable energy use, promotes true cost accounting and a steady state or slow growth economy (Caradonna, 2014; Niesenbaum, 2020). A sustainable environment maintains and preserves resources, as it protects biodiversity and ecosystem services. Social equality or equity maintains, among other elements, that all humans have equal opportunities such as equal distribution of resources, fair wages, democratic institutions, social justice, the ability to live a safe and meaningful life and rising standards of living for all members (Caradonna, 2014).

Sustainability is achieved at the intersection of these three pillars (Caradonna, 2014; Niesenbaum, 2020). For example, ecological economies that incentivize the development of renewable energy technologies benefit both the environment and equity, as fossil fuel consumption will decrease and job opportunities will increase, respectively (Niesenbaum, 2020). Such increased job opportunities benefit the environment and social equity, as natural resources will be conserved and wages will increase the standard of living of individuals, respectively. Lastly, improving environmental conditions by way of supporting renewable energy technologies can improve upon social equity and the economy. For instance, often those in less developed nations are bearing the brunt of climate change (Niesenbaum, 2020). Thus, if environmental conditions improve, all people will have more equitable access to the earth's resources. Moreover, conserving resources ensures economic longevity, as there will be resources available for the next generation.

While the aforementioned model stresses three pillars of sustainability, some contend that a fourth dimension should be added: esthetics. Esthetics is not just what is visually appealing, but sensory perception in general (Allen, 2019). From a young age, we perceive our immediate surroundings through all of our senses and these perceptions influence our values and judgments (Allen, 2020), as well as our

"...preferences, phobias, and behaviors." Such development often occurs in the context of how we experience a specific place (Orr, 2010, p. 152). As such, we are motivated to protect those places that reflect our preferences and values (Allen, 2020; Orr, 2010). Moreover, esthetics can be used to bring attention to communicate and educate people on issues related to sustainability, such as using writing and art to communicate the dangers of climate change or to promote recycling (Allen, 2019). Thus, esthetics serves two purposes: to educate and bring awareness to matters of sustainability and to develop our values, judgments, and preferences.

A solution that results in the intersection of all four sustainability pillars is supporting local economies (Orr, 2010). Global economies have resulted in income inequality (Niesenbaum, 2020) and the decay of incomes for many as markets seek cheaper labor (Orr, 2010). Such decay has decreased social equity as many struggle to access such vital needs as healthcare, housing, and food (Orr, 2010). By supporting local economies, money circulates in the locality, and then creates a "...multiplier effect whereby local jobs and investment occur" (Orr, 2010, p. 157). Environmentally, goods are not transported long distances, thereby reducing the environmental impact of production. Esthetically, consumers who shop locally find value in their locality, and thus want to protect it from environmental, economic and equity degradation (Orr, 2010).

2.1.6 Systems Theory

As noted previously, sustainability can best be achieved by utilizing a systems thinking, or a systems theory approach. A system is "...an assemblage of objects united by some form of regular interaction or interdependence" (Mele et al., 2010, p. 129). Thus, systems exist all around us. For instance, cells, organisms, and societies are systems (Laszlo & Krippner, 1998), as is the universe (Mele et al., 2010).

Systems theory involves identifying all of the parts that comprise a system, determining how those parts are interrelated, and how outside factors influence those relationships and the overall phenomena itself (Kim, 1999; Littlejohn & Foss, 2009). Mele et al. (2010) define systems theory as:

> "...a theoretical perspective that analyzes a phenomenon seen as a whole and not as simply the sum of elementary parts. The focus is on the interactions and on the relationships between parts in order to understand an entity's organization, functioning and outcomes. This perspective implies a dialogue between holism and reductionism" (p. 127).

While the establishment of systems theory or systems thinking is generally attributed to Von Bertalanffy's General Systems Theory (Hammond, 2010; Laszlo & Krippner, 1998), systems thinking can be traced to sixth century Greek philosophy as philosophers generally attempted to make sense of the chaotic world around them and specifically to Aristotle's third century B.C. metaphysics which emphasized the concept that the whole is greater than the sum of its parts (Von Bertalanffy, 1972).

Systems theory was developed to identify universal laws common to all systems, and therefore it is applicable to various fields of study (Mele et al., 2010) including fashion (Sinha et al., 2016). Within the context of apparel and textiles, systems theory has been used to illustrate the intersection of textiles, apparel, and sustainability (Hawley, 2006; Kennedy & Parsons, 2011; Kennedy et al., 2018; Kozlowski et al., 2018; Moore & Ausley, 2004). Common themes across systems theory employed in apparel, textile, and sustainability studies include an emphasis in involving all stakeholders so as to fully understand phenomena, as well as understanding the relationships among the parts that contributes to a more comprehensive understanding of the whole system. By analyzing each part of a system, an understanding of how the parts operate independently, as well as how each part can influence another part by identifying positive and negative feedback loops can be developed (Niesenbaum, 2020). Moreover, this holistic perspective allows for the identification of leverage points, or points of intervention that might have the most significant impact in changing the system (Meadows, 1999).Indeed, a leverage point to promote sustainability is to reduce consumer consumption. The capitalist nature of current consumption patterns is unsustainable, as it results in our society living outside of its ecological limits, as well as increases social inequity (Caradonna, 2014). While it may be difficult to reduce consumption entirely, an alternative, influencing SRFC, could help to reduce the negative effects of consumption by shifting the current unsustainable nature of the fashion industry to one that engages in sustainable production and consumption.

2.2 Method

2.2.1 Conceptual Model

Systems theory uses loops to demonstrate how a system is connected (Kim, 1999). The basic mechanics of the loops include the name of the loop located in its center, which demonstrates what is being impacted. Arrows are included to communicate that one variable affects the next one either in a positive or negative direction, indicated by a + or -, respectively. The type of loop can either be a reinforcing or a balancing loop, indicated by R or B, respectively (Kim, 1999). Reinforcing loops occur when changing a variable in one direction produces a response in the same direction of the next variable (Kim, 1999). Balancing loops occur when there are an odd number of opposite links, for instance two positives and one negative (Kim, 1999). As such, balancing loops counteract change (Kim, 1999).

Using systems theory, the following conceptual model demonstrates two interconnected systems: sustainability and SRFC (see Fig. 2.1). The central system of sustainability, depicted within the diamond of Fig. 2.1, illustrates a reinforcing feedback loop among key pillars: esthetics, equity, environment, and economy. For instance, supporting local economies enhances equity by providing more job

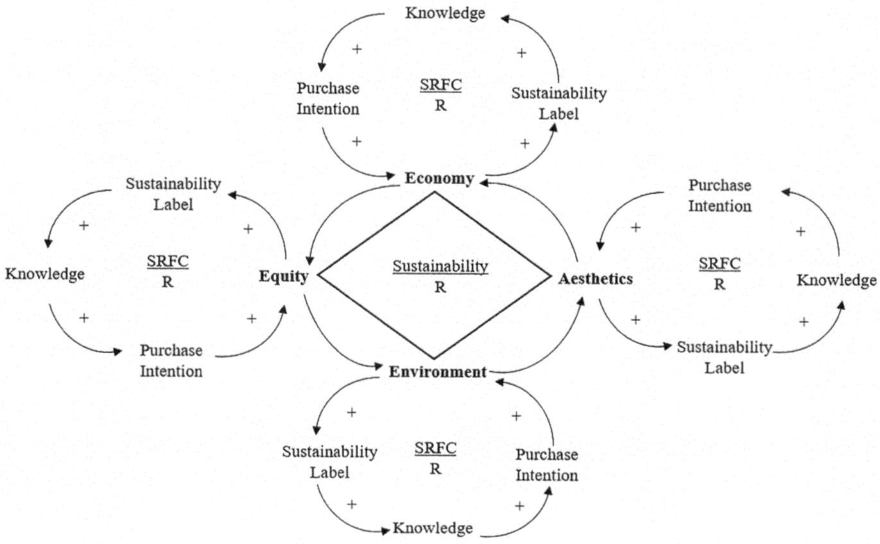

Fig. 2.1 Sustainability Conceptual Model

opportunities and income access. Increased equity leads to environmental benefits, as people are less inclined to exploit the environment for economic gain. Environmental support, in turn, fosters the esthetic value of sustainability, as people appreciate and safeguard the environment. Additionally, esthetic tools like labeling aid in communicating sustainable initiatives to consumers, thus fostering an economy aligned with sustainable values. This positive feedback loop strengthens and expands sustainability as each pillar of sustainability is supported.

The second system, SRFC, is demonstrated in Fig. 2.1 as separate, reinforcing loops that positively act upon each of the pillars of sustainability. For instance, sustainability labeling positively increases consumer knowledge, which positively impacts purchase intention of sustainably-labeled products by providing the necessary information for consumers to make informed purchase decisions (Blazquez et al., 2020; Cowan & Kinley, 2014; Diddi & Niehm, 2016; Gallastegui, 2002; Goworek et al., 2012; Hyllegard et al., 2012; Kang et al., 2013; Ko & Jin, 2017; Kozar & Connell, 2013; Okur & Saricam, 2018; Zheng & Chi, 2015). By purchasing sustainability-labeled goods, each foundation of sustainability is supported, as the consumption of sustainable apparel, as opposed to conventionally-produced apparel, reduces the negative impact on the environment and equity, supports an ecological economy, and enables consumers to consume in a manner consistent with their values of sustainability.

2.2.2 *Case Study*

To evaluate the conceptual model, The Standard Clothing Company (a pseudonym) was chosen, as it is a locally-based small apparel retail brand that utilizes sustainability labeling to communicate both the environmental and social impacts of its apparel production. The Standard Clothing Company's website and social media were mined, and text analyzed for information pertaining to their sustainability labeling program. Additionally, an interview was conducted with The Standard Clothing Company's brand manager to better understand the company's sustainability labeling strategy. This interview was recorded and transcribed verbatim. Data sampling and collection reached saturation, that is, once the same information began to emerge in the data collection, the data collected was considered sufficient for understanding the phenomena (Hodges, 2011; Lincoln & Guba, 1985). The interview transcript, along with the social media and website data were analyzed for recurring themes (Spiggle, 1994). Four themes were used to guide the interpretation of results: the economic, environmental, equitable, and esthetic value of sustainability labels.

2.3 Results

2.3.1 *The Standard Clothing Company*

Located in the state of North Carolina, The Standard Clothing Company focuses on rebuilding the American culture of making things by selling premium quality t-shirts that are grown and sewn in the USA. The company is interested in reinventing the fashion industry by changing the way in which consumers and the industry itself connects with the supply chain of apparel. As such, The Standard Clothing Company's "dirt to shirt" concept provides consumers with complete transparency of the supply chain. For instance, cotton for the t-shirts is grown in North Carolina by area farmers, it is then loomed into yarn by a nearby mill, and then knitted into fabric, cut and sewn and dyed by local factories.

Because The Standard Clothing Company utilizes practices that minimize waste and maximize social and environmental benefits, the t-shirt it produces is a sustainable fashion item as defined by Kozlowski et al. (2018). Moreover, The Standard Clothing Company's t-shirt is considered sustainable based on Niesenbaum's (2020) definition because the product enables the current generation to meet their own needs without compromising the ability of future generations to do the same. It also provides economic development and equitable economic opportunity, and therefore provides an elevated standard of living.

2.3.2 The Standard Clothing Company and Sustainability Labeling

Unlike many brands that keep their supply chains private, The Standard Clothing Company invites its consumers to learn more about the apparel production process by attaching sustainability labels to their t-shirts. Each label features a unique QR code that, when scanned, allows consumers to see how and where their shirt was made. This label enables consumers to gain knowledge concerning the contractors and production processes involved along The Standard Clothing Company's supply chain. By providing such transparency, The Standard Clothing Company hopes to educate its consumers regarding the intricacies of the apparel supply chain and the effects of sustainable sourcing, and in turn, to influence consumers to purchase sustainably-produced apparel. As such, The Standard Clothing Company supports both sustainability and SRFC by way of sustainability labeling as explained below.

2.3.3 Economy

Figure 2.2 outlines how The Standard Clothing Company's sustainability labeling supports the sustainability foundation of economy. According to Fig. 2.2, beginning at the bottom of the loop with economy and moving counterclockwise, sustainability labeling positively influences consumer knowledge and subsequently the purchase intention of sustainably-labeled clothing and therefore positively influences SRFC. Specifically, The Standard Clothing Company supports its mission of rebuilding the American culture of making things by contracting with local vendors, as every contractor in their supply chain, from cotton farmer to yarn spinner, to fabric producer to cut and sew operator is locally based, there is no outsourcing of any steps of the supply chain. As a result, local economies are financially strengthened when t-shirts are purchased. This money can then circulate throughout the local economy, leading to job security for current employees and potential job opportunities for future ones. Furthermore, by using a natural resource that is sustainably farmed to produce its shirts, and production practices that minimize waste, The Standard Clothing Company's production practices respect ecological limits.

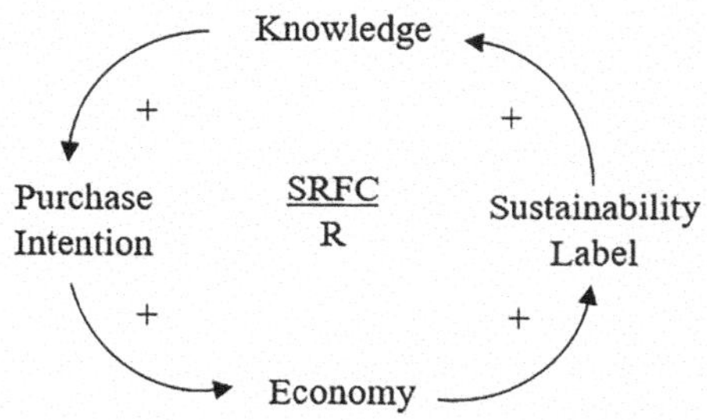

Fig. 2.2 Economy Loop

Lastly, by selling a high-quality t-shirt that lasts longer, The Standard Clothing Company minimizes consumption of additional t-shirts. The sustainable benefits of the Standard Clothing Company's t-shirt are made known to consumers via their sustainability labeling, thereby improving consumer knowledge regarding the sustainability of their t-shirts. Conceptually, this knowledge then positively influences consumer's purchase intention of The Standard Clothing Company's t-shirts, resulting in SRFC.

2.3.4 *Environment*

Figure 2.3 outlines how sustainability labeling positively influences the foundation of the environment, beginning at the top of the loop with environment and moving counterclockwise, as The Standard Clothing Company utilizes sustainability labeling to positively influence consumer knowledge regarding their environmentally friendly production methods, which, in turn, positively influences consumer's purchase intention of The Standard Clothing Company's sustainable t-shirts, therefore positively supporting SRFC. Particularly, The Standard Clothing Company's sustainability labeling allows consumers to become knowledgeable about their production processes, including that The Standard Clothing Company's use of a local supply chain reduces the environmental cost of shipping and transporting compared with overseas production, which requires the environmentally taxing forms of container ships or air transportation for shipping. Additionally, local production reduces harmful environmental effects, such as polluting local water sources via toxic dyeing processes, as there are more rigid environmental protection laws in the USA than in many of the developing countries that currently produce fashion products. For instance, The Standard Clothing Company partners with the local water authority to ensure the dyes being used in the production of their clothing can be properly cleaned and managed.

Furthermore, many of the contractors that The Standard Clothing Company partners with also implement environmentally friendly processes, such as zero-waste design and regenerative agricultural practices, including cover cropping and no-till planting. As a result, the environment is less degraded than happens with conventional apparel production practices. Moreover, The Standard Clothing Company's

Fig. 2.3 Environment Loop

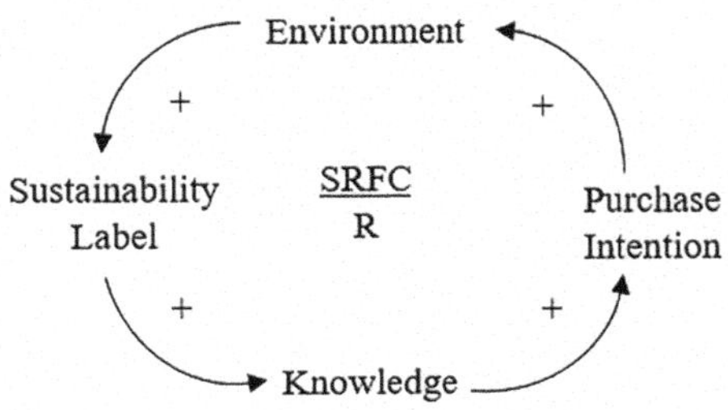

t-shirt is made from 100% locally grown cotton, a natural resource, unlike the popular oil-based polyester fabric that is common in many apparel products. Lastly, The Standard Clothing Company's t-shirts are made to last, as they are reinforced in the areas that experience the most wear and tear. As such, their products benefit the environment, as these t-shirts are not readily disposed of, and can be disposed of via composting. Thus, according to Fig. 2.3, by providing sustainability labeling, consumers are more knowledgeable regarding the aforementioned environmental effects of the Standard Clothing Company's production methods. Such knowledge positively influences consumer's purchase intention of sustainably-labeled t-shirts, therefore reinforcing SRFC.

2.3.5 *Equity*

Figure 2.4 details the positive effect of The Standard Clothing Company's sustainability labeling on the foundation of equity, beginning on the right-hand side of the figure and moving counterclockwise, as The Standard Clothing Company utilizes sustainability labeling to communicate the equitable impacts of their production methods, which positively influences consumer knowledge regarding production of their t-shirts, and subsequently consumer's purchase intention of these t-shirts. For instance, The Standard Clothing Company's sustainability labeling allows consumers to realize that the company partners with local cotton farmers, a profession that is subject to commodity pricing and has been driven out due to the commercialization of agriculture. By partnering with local farmers, and guaranteeing them a fair price for their products, farmers are better able to maintain and grow their businesses. Moreover, The Standard Clothing Company partners with local contractors, many of whom support social responsibility and are worker-owned, thereby improving social equity. Lastly, The Standard Clothing Company's proximity to their supply chain enables them to forge trusting relationships. Therefore, they can hold their contractors accountable to ensure that each step along their supply chain is environmentally and socially responsible. Thus, according to Fig. 2.4, as a result of sustainability labeling, consumers are more knowledgeable regarding how The Standard

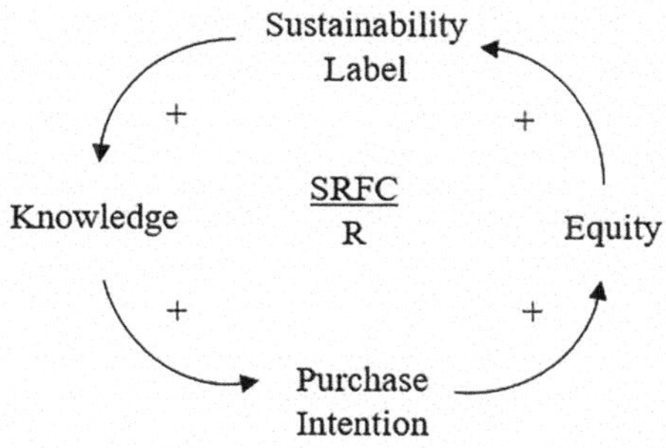

Fig. 2.4 Equity Loop

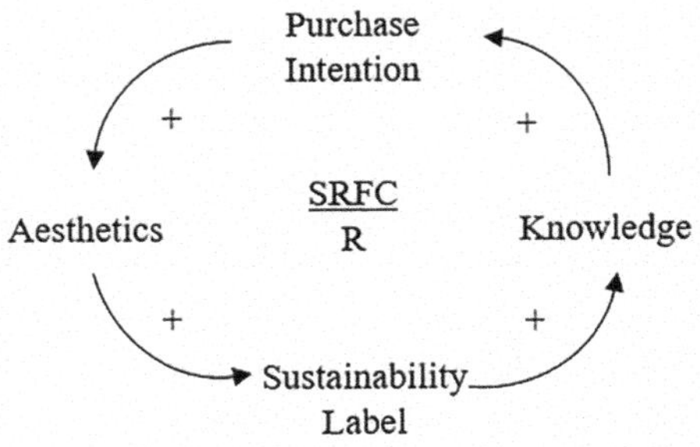

Fig. 2.5 Esthetics Loop

Clothing Company's production methods support equity, thus they are positively influenced to purchase sustainably-labeled t-shirts, which reinforces SRFC.

2.3.6 Esthetics

Finally, Fig. 2.5 outlines the effects of sustainability labeling on the foundation of esthetics, beginning on the left of the figure and moving counterclockwise, as The Standard Clothing Company utilizes their sustainability labeling to enable consumers to act on values of sustainability. For instance, as previously noted, The Standard Clothing Company's supply chain vendors implement sustainable farming techniques, support social equity, and invest in local economies which ensures environmental, social, and economic sustainability. The Standard Clothing Company's sustainability labeling makes such efforts known, thereby positively influencing consumers' knowledge of such processes. This knowledge positively influences consumers to purchase The Standard Clothing Company's t-shirt, thereby enabling consumers to purchase products in alignment with their sustainability values.

2.4 Conclusion

The retail industry is experiencing reinvention as it reorganizes itself to better meet consumer needs (Wang, 2020). Instead of implementing experiential aspects of retail, a popular, current trend (Wang, 2020), fashion retailers can reinvent themselves by using sustainability labeling on the apparel they sell to become the link between apparel producers and consumers (James & Montgomery, 2017b; Lehner, 2014; Vadakkepatt et al., 2021). Such labeling will address the consumer's desire for apparel production transparency (Amed et al., 2019; Ditty, 2020; Goworek et al., 2012; Sustainable Apparel Coalition, 2019; Williams & Hodges, 2022b; Vehmas et al., 2018; Yudina, 2017) and interest in consuming sustainably (Cone Communications, 2017; Das et al., 2023; Fiedler et al., 2020). Retailers who adopt such an initiative will become trusted destinations for facilitating consumer access to the transparency of their supply chains (Sides & Skelly, 2021), and will benefit from increased consumer trust (Sustainable Apparel Coalition, 2019). While previous research has investigated apparel labeling (Baker, 2002; Hilowitz, 1997; Koszewska, 2011) and its effect on attitude toward the brand and purchase intention of the product (D'Souza et al., 2006; Dickson, 2001; Hustvedt et al., 2008; Hyllegard et al., 2012) there is little research that investigates the use of sustainability labeling to meet the consumer's apparel sustainability needs. Therefore, the purpose of this study was to develop a conceptual model based on systems theory to demonstrate how retailers can utilize sustainability labeling to appeal to the consumer's interest in accessing apparel sustainability information and desire to consume apparel more sustainably.

Sustainability results when the following four pillars of sustainability intersect: esthetics, economy, environment, and social equity (Allen, 2019; Caradonna, 2014; Niesenbaum, 2020). Investigating sustainability labeling using systems theory illuminates the interdependent nature of the pillars of sustainability, and how fashion retailers can reinvent themselves by instituting sustainability labeling to meet consumer's interest in supporting sustainable apparel consumption. While this model outlines the interconnected relationship among the pillars of sustainability, as well as the relationship among sustainability, sustainability labeling, consumer knowledge, purchase intention, and SRFC, there are other variables that may contribute to SRFC that are not included in this model, such as consumer values and motivations. Moreover, while this model posits that retailers can promote SRFC, and more broadly sustainability, this model does not singularly answer consumers' calls for more transparency in the fashion industry (Sustainable Apparel Coalition, 2019), as brands must first be willing to disclose such information. However, this model does suggest that retailers can play a significant role in meeting consumer interest for apparel production information, and desire in consuming sustainably, by including sustainability labeling on the apparel products they sell.

For instance, esthetically, sustainability labels can be used as tools to attract consumer attention and educate them on issues of sustainability, as well as a mechanism that enables consumers to consume apparel in accordance with their values of sustainability. Specifically, in the case of The Standard Clothing Company, sustainability labeling enables consumers to track the entire supply chain of their t-shirt products. As such, consumers become educated regarding how, where and who made their shirt. Moreover, these esthetic tools allow consumers to engage in SRFC, and thus consume in a manner that is aligned with their values of sustainability. Therefore, by offering sustainability labeling, retailers can facilitate consumers' access to information related to the sustainability of their apparel, which influences SRFC, thereby meeting consumer's interests in transparency and engaging in sustainable apparel consumption.

Economically, The Standard Clothing Company's sustainability labeling supports ecological economics by enabling consumers to support local economies because The Standard Clothing Company utilizes and is transparent about using local vendors in the production of their t-shirts. The purchase of a t-shirt from The Standard Clothing Company can therefore strengthen local economies by supporting local jobs. Furthermore, by utilizing a local supply chain, The Standard Clothing Company respects ecological limits. The sustainable, economic effect of engaging in SRFC is made aware to consumers via sustainability labeling. Therefore, the label allows for consumers to access desired transparency, which influences SRFC, thereby simultaneously supporting the pillars of sustainability and meeting consumer's apparel sustainability needs.

In terms of social equity, The Standard Clothing Company itself is dedicated to paying fair rates to their contractors, many of which are employee-owned and or support social responsibility. As a result, the standard of living improves for these individuals as they are paid fair wages and have a voice in the operation of their company. By using sustainability labeling, and communicating their supply chain to consumers, The Standard Clothing Company allows its consumers to understand how their purchases equitably support others. As such, sustainability labeling offers retailers the opportunity to fulfill consumer desire for apparel supply chain transparency, giving them the knowledge necessary to participate in SRFC.

Environmentally, The Standard Clothing Company supports sustainability by communicating to its customers the environmental impact of their products via sustainability labeling. By using sustainability labeling, consumers better understand how the local nature of The Standard Clothing Company's shirt production reduces the environmental cost of apparel production. Moreover, by partnering with local contractors who support environmentally friendly production practices, as well as using cotton, a natural, renewable resource, for its t-shirts, The Standard Clothing Company minimizes the impact of its production on the environment. By using sustainability labeling, retailers can increase consumers' awareness of the environmental benefits of their production, thus fulfilling their interests in engaging in SRFC and supporting sustainability.

2.4.1 Limitations and Future Research

This study uses a case study approach to evaluate the developed conceptual model. Consequently, the results can only be generalized to companies like The Standard Clothing Company. Although The Standard Clothing Company is considered a sustainable apparel retailer, previous research demonstrates that fashion retailers in general can benefit from reporting the transparency of their supply chains, even the negative aspects, as long as they articulate how and why they plan to change such aspects (Sustainable Apparel Coalition, 2019). Nevertheless, additional research should explore the ways in which sustainability labeling affects retailers with unsustainable supply chains. Such a study could determine whether and how retailers with such supply chains benefit from sustainability labeling.

Additionally, the conceptual model suggests that sustainability labeling positively influences consumer knowledge, which positively influences purchase intention of sustainably-labeled apparel, as this is what the literature suggests. However, this relationship was not empirically tested, and thus conclusions regarding the success of sustainability labeling supporting sustainability rest on whether sustainability labeling positively influences consumer knowledge and if such knowledge positively affects consumer's purchase intentions. Thus, future research should empirically test the conceptual model using additional sustainability labeling programs to determine their effect on sustainability. For instance, a study that utilizes a survey or experiment to test the effect of sustainability labeling on consumer knowledge, and the effect of knowledge on purchase intention could be pursued. Such a study could determine the influence of sustainability labeling on consumer knowledge, and the effect of consumer knowledge on consumer's purchase intention of sustainably-labeled apparel. Moreover, an exploratory study that investigates the factors that moderate consumer knowledge and purchase intention should also be performed to better understand how consumers process and utilize information on sustainability labeling. Such research endeavors can support the role of retailers as the link between apparel producers and consumers, a role which will enable retailers to reinvent themselves by answering the consumer's calls for more sustainability through transparency.

References

Alldredge, K., & Teichner, W. (2023). *For US consumers, it's a matter of 'and'-not 'or'*. McKinsey & Company. https://www.mckinsey.com/industries/consumer-packaged-goods/our-insights/for-us-consumers-its-a-matter-of-and-not-or#/

Allen, A. (2019). Aesthetics and sustainability. In W. Filho (Ed.), *Encyclopedia of sustainability in higher education* (pp. 1–9). Springer. https://doi.org/10.1007/978-3-319-63951-2

Allen, A. (2020). Sounding sustainable; or, the challenge of sustainability. In T. J. Cooley (Ed.), *Cultural sustainabilities: Music, media, language, advocacy* (pp. 43–59). University of Illinois Press. https://doi.org/10.5406/j.ctvh9w1f9

Amed, I., Berg, A., Balchandani, A., Andersson, J., Hedrich, S., & Young, R. (2019). *The state of fashion 2019*. McKinsey & Company. https://www.mckinsey.com

Baker, S. (2002). *Hang tags: Connecting to consumers*. https://www.just-style.com/analysis/connecting-to-consumers_id93012.aspx

Berg, A., et al. (2020). *Fashion on climate: How the fashion industry can urgently act to reduce its greenhouse gas emissions*. McKinsey & Company and Global Fashion Agenda. https://www.mckinsey.com/~/media/mckinsey/ industries/retail/ our%20insights/fashion%20on%20climate/fashion-on-climate-full-report.pdf

Bernard, J. C., Hustvedt, G., & Carroll, K. A. (2013). What is a label worth? Defining the alternatives to organic for US wool producers. *Journal of Fashion Marketing and Management, 7*(5), 266–279. https://doi.org/10.1108/JFMM-01-2013-0009

Blazquez, M., Henninger, C., Alexander, B., & Franquesa, C. (2020). Consumers' knowledge and intentions towards sustainability: A Spanish fashion perspective. *Fashion Practice, 12*(1), 34–54. https://doi.org/10.1080/17569370.2019.1669326

Bluesign. (n.d.). https://www.bluesign.com/en

Brosdahl, D. J. C., & Carpenter, J. M. (2010). Consumer knowledge of the environmental impact of textile and apparel production, concern for the environment, and environmentally friendly consumer behavior. *Journal of Textile and Apparel Technology and Management, 6*(4), 1–9.

Byrd, K., & Su, J. (2021). Investigating consumer behaviour for environmental, sustainable and social apparel. *International Journal of Clothing Science and Technology, 33*(3), 336–352. https://doi.org/10.1108/IJCST-03-2020-0040

Caradonna, J. (2014). *Sustainability: A history*. Oxford University Press.

Care and Fair. (n.d.). https://www.care-fair.org/en/start-english/

Chrobot, P., Faist, M., Gustavus, L., Martin, A., Stamm, A., Zah, R., & Zollinger, M. (2018). *Measuring fashion: Environmental impact of the global apparel and footwear industries.* Quantis. https://quantis.com/wp-content/uploads/2018/03/measuringfashion_ globalimpact-study_full-report_quantis_cwf_2018a.pdf

Clean Clothes Campaign. (n.d.). https://cleanclothes.org/

Cone Communications. (2017). *2017 Cone Gen Z CSR study: How to speak Z.* Cone Communications. https://www.conecomm.com/research-blog/2017-genz-csr-study

Connell, K. Y. H. (2010). Internal and external barriers to eco-conscious apparel acquisition. *International Journal of Consumer Studies, 34*, 279–286. https://doi.org/10.1111/j.1470-6431.2010.00865.x

Connell, K. Y. H., & Kozar, J. M. (2012). Sustainability knowledge and behaviors of apparel and textile undergraduates. *International Journal of Sustainability in Higher Education, 13*(4), 394–407. https://doi.org/10.1108/14676371211262335

Cowan, K., & Kinley, T. (2014). Green spirit: Consumer empathies for green apparel. *International Journal of Consumer Studies, 38*(5), 493–499. https://doi.org/10.1111/ijcs.12125

Cradle to Cradle. (n.d.). https://www.c2ccertified.org/

D'Souza, C. (2000). Bridging the communication gap: Dolphin-safe 'ecolabels'. *Corporate Communications, 5*(4), 185–189. https://doi.org/10.1108/13563280010379129

D'Souza, C., Taghian, M., & Lamb, P. (2006). An empirical study on the influence of environmental labels on consumers. *Corporate Communications, 11*(2), 162–173. https://doi.org/10.1108/13563280610661697

Das, R., Kalia, S., & Kuijpers, D. (2023). *Zero consumers': What they want and why it matters.* McKinsey & Company. https://www.mckinsey.com/~/media/mckinsey/industries/retail/our%20insights/zero%20consumers%20what%20they%20want%20and%20why%20it%20matters/zero-consumers-what-they-want-and-why-it-matters.pdf?shouldIndex=false

Dickson, M. A. (2001). Utility of no sweat label of apparel consumers: Profiling label users and predicting their purchases. *The Journal of Consumer Affairs, 35*(1), 96–119. https://www.jstor.org/stable/23860073

Diddi, S., & Niehm, L. S. (2016). Corporate social responsibility in the retail apparel context: Exploring consumers' personal and normative influences on patronage intentions. *Journal of Marketing Channels, 23*(1–2), 60–76. https://doi.org/10.1080/1046669X.2016.1147892

Ditty, S. (2020). *Consumer survey key findings*. Fashion Revolution. https://issuu.com/fashionrevolution/docs/fashrev_consumersurvey_2020_keyfindings

Fair Trade. (n.d.). https://www.fairtrade.net/

Fair Wear. (n.d.). https://www.fairwear.org/

Federal Trade Commission. (n.d.). *Threading your way through the labeling requirements under the textile and wool acts.* https://www.ftc.gov/tips-advice/business-center/guidance/threading-your-way-through-labeling-requirements-under-textile#place

Feltham, T. S., & Martin, L. A. (2006). *Apparel care labels: Understanding consumers' use of information*. [Abstract]. ASAC. https://ssrn.com/abstract=1809865

Fiedler, L., Hazan, E., Ruwadi, B., & Ungerman, K. (2020). *Retail reimagined: The new era for customer experience*. https://www.mckinsey.com/~/media/McKinsey/Business%20Functions/Marketing%20and%20Sales/Periscope/Insights/Survey%20Reports/Reinventing%20Retail/Periscopes%20Retail%20Reimagined%20Report%202020-August-2020.pdf.

Gallastegui, I. (2002). The use of eco-labels: A review of the literature. *European Environment, 12*, 316–331. https://doi.org/10.1002/eet.304

Gil-Saura, I., Ruiz-Molina, M.-E., Marín-García, A., & Michel, G. (2023). Sustainability-oriented commerce innovation: How does it influence consumer satisfaction? *International*

Journal of Retail & Distribution Management, 51(11), 1477–1496. https://doi.org/10.1108/IJRDM-07-2022-0267

Global Organic Textile Standard. (n.d.). https://global-standard.org/

Goworek, H., Fisher, T., Cooper, T., Woodward, S., & Hiller, A. (2012). The sustainable clothing market: An evaluation of potential strategies for UK retailers. *International Journal of Retail & Distribution Management, 40*(12), 935–955. https://doi.org/10.1108/09590551211274937

Gupta, S., & Gentry, J. G. (2018). Evaluating fast fashion: Examining its micro and the macro perspective. In C. Becker-Leifhold & M. Heuer (Eds.), *Eco-friendly and fair: Fast fashion and consumer behavior* (pp. 15–23). Routledge.

Hammond, D. (2010). *The science of synthesis: Exploring the social implications of general systems theory*. University Press of Colorado.

Han, J., Seo, Y., & Ko, E. (2017a). Staging luxury experiences for understanding sustainable fashion consumption: A balance theory application. *Journal of Business Research, 74*, 162–167. https://doi.org/10.1016/j.jbusres.2016.10.029

Han, S. L.-C., Henninger, C. E., Apeagyei, P., & Tyler, D. (2017b). Determining effective sustainable fashion communication strategies. In C. Henninger, P. Alevizou, H. Goworek, & D. Ryding (Eds.), *Sustainability in fashion* (pp. 127–149). Springer. https://doi.org/10.1007/978-3-319-51253-2

Harris, F., Roby, H., & Dibb, S. (2016). Sustainable clothing: Challenges, barriers and interventions for encouraging more sustainable consumer behaviour. *International Journal of Consumer Studies, 40*(3), 309–318. https://doi.org/10.1111/ijcs.12257

Hawley, J. M. (2006). Digging for diamonds: A conceptual framework for understanding reclaimed textile products. *Clothing & Textiles Research Journal, 24*(3), 262–275. https://doi.org/10.1177/0887302X062946

Hill, J., & Lee, H.-H. (2012). Young Generation Y consumers' perceptions of sustainability in the apparel industry. *Journal of Fashion Marketing and Management, 16*(4), 477–491. https://doi.org/10.1108/13612021211265863

Hilowitz, J. (1997). Social labelling to combat child labour: Some considerations. *International Labour Review, 136*, 215–232.

Hodges, N. (2011). Qualitative research: A discussion of frequently articulated qualms (FAQs). *Family and Consumer Sciences Research Journal, 40*(1), 90. https://doi.org/10.1111/j.1552-3934.2011.02091.x

Hofenk, D., van Birgelen, M., Bloemer, J., & Semijn, J. (2019). How and when retailers' sustainability efforts translate into positive consumer responses: The interplay between personal and social factors. *Journal of Business Ethics, 156*, 473–492. https://doi.org/10.1007/s10551-017-3616-1

Hustvedt, G., & Bernard, J. C. (2008). Consumer willingness to pay for sustainable apparel: The influence of labelling for fibre origin and production methods. *International Journal of Consumer Studies, 32*, 491–498. https://doi.org/10.1111/j.1470-6431.2008.00706.x

Hustvedt, G., & Bernard, J. C. (2010). Effects of social responsibility labeling and brand willingness to pay for apparel. *International Journal of Consumer Studies, 34*, 619–626. https://doi.org/10.1111/j.1470-6431.2010.00870.x

Hustvedt, G., Peterson, H. H., & Chen, Y.-J. (2008). Labelling wool products for animal welfare and environmental impact. *International Journal of Consumer Studies, 32*, 427–437.

Hyllegard, K. H., Yan, R.-N., Ogle, J. P., & Lee, K.-H. (2012). Socially responsible labeling: The impact of hang tags on consumers' attitudes and patronage intentions toward an apparel brand. *Clothing and Textiles Research Journal, 30*(1), 51–66. https://doi.org/10.1177/0887302X11433500

Hyllegard, K. H., Ogle, J. P., & Yan, R. (2014). College students' responses to prosocial marketing claims on apparel hang tags. *Journal of Fashion Marketing and Management, 18*(3), 269–283. https://doi.org/10.1108/JFMM-12-2012-0075

Intergovernmental Panel on Climate Change. (2014). *Climate change 2014: Mitigation of climate change*. https://www.ipcc.ch/report/ar5/wg3/

International Organization for Standardization. (2006). *ISO 14040:2006(en)-environmental management-life cycle assessment-principles and framework*. https://www.iso.org/obp/ui/#iso:std:iso:14040:ed-2:v1:en

International Organization for Standardization. (2010). *ISO 26000: 2010(en): Guidance on social responsibility*. https://www.iso.org/obp/ui/#iso:std:iso:26000:ed-1:v1:en

Iwanow, H., McEachern, M. G., & Jeffrey, A. (2005). The influence of ethical trading policies on consumer apparel purchase decisions: A focus on The Gap Inc. *International Journal of Retail & Distribution Management, 33*(5), 371–387. https://doi.org/10.1108/09590550510596740

Jäger, A.-K., & Weber, A. (2020). Increasing sustainable consumption: Message framing and in-store technology. *International Journal of Retail & Distribution Management, 48*(8), 803–824. https://doi.org/10.1108/IJRDM-02-2019-0044

James, A., & Montgomery, B. (2017a). Connectivity, understanding and empathy: How a lack of consumer knowledge of the fashion supply chain is influencing socially responsible fashion purchasing. In S. Muthu (Ed.), *Textiles and clothing sustainability: Sustainable fashion and consumption* (pp. 61–95). Springer. https://doi.org/10.1007/978-981-10-2131-2

James, A., & Montgomery, B. (2017b). The role of the retailer in socially responsible fashion purchasing. In S. Muthu (Ed.), *Textiles and clothing sustainability: Sustainable fashion and consumption* (pp. 1–40). Springer. https://doi.org/10.1007/978-981-10-2131-2

Kang, J., Liu, C., & Kim, S.-H. (2013). Environmentally sustainable textile and apparel consumption: The role of consumer knowledge, perceived consumer effectiveness and perceived personal relevance. *International Journal of Consumer Studies, 37*, 442–452. https://doi.org/10.1111/ijcs.12013

Kennedy, A.-M., & Parsons, A. (2011). Macro-social marketing and social engineering: A systems approach. *Journal of Social Marketing, 2*(1), 37–51. https://doi.org/10.1108/20426761211203247

Kennedy, A.-M., Kapitan, S., Bajaj, N., Bakonyi, A., & Sands, S. (2018). Case study 18: Fast fashion: A wicked problem for macro-social marketing. In G. Hastings & C. Domegan (Eds.), *Social marketing: Rebels with a cause* (pp. 502–508). Routledge.

Kim, D. (1999). Introduction to systems thinking. *Pegasus Communications*.

Kim, H.-S., & Damhorst, M. L. (1998). Environmental concern and apparel consumption. *Clothing and Textiles Research Journal, 16*(3), 126–133. https://doi.org/10.1177/0887302X9801600303

Ko, S. B., & Jin, B. (2017). Predictors of purchase intention toward green apparel products. *Journal of Fashion Marketing and Management, 21*(1), 70–87. https://doi.org/10.1108/JFMM-07-2014-0057

Koszewska, M. (2011). Social and eco-labelling of textile and clothing goods as a means of communication and product differentiation. *Fibres and Textiles in Eastern Europe, 19*(4), 20–26.

Koszewska, M. (2015). Life cycle assessment and the environmental and social labels in the textile and clothing industry. In S. Muthu (Ed.), *Handbook of life cycle assessment (LCA) of textiles and clothing* (pp. 325–344). Woodhead Publishing. https://doi.org/10.1016/B978-0-08-100169-1.00015-0

Kozar, J. M., & Connell, K. Y. H. (2010). Socially responsible knowledge and behaviors: Comparing upper- vs. lower- classmen. *College Student Journal, 44*(2), 279–293. https://krex.k-state.edu/dspace/bitstream/handle/2097/8376/HillerConnellCSJ2010.pdf ?sequence=1&isAllowed=y

Kozar, J. M., & Connell, K. Y. (2013). Socially and environmentally responsible apparel consumption: Knowledge, attitudes, and behaviors. *Social Responsibility Journal, 9*(2), 316–325. https://doi.org/10.1108/SRJ-09-2011-0076

Kozlowski, A., Searcy, C., & Bardecki, M. (2018). The redesign canvas: Fashion design as a tool for sustainability. *Journal of Cleaner Production, 183*, 194–207. https://doi.org/10.1016/j.jclepro.2018.02.014

Label Consciousness. (2007, April 26). *Women's Wear Daily, 193*(88), 2. EBSCO.

Laitala, K., & Klepp, I. G. (2013). Environmental and ethical perceptions related to clothing labels among Norwegian consumers. *Research Journal of Textile and Apparel, 17*(1), 50–58. https://doi.org/10.1108/RJTA-17-01-2013-B005

Laszlo, A., & Krippner, S. (1998). Systems theories: Origins, foundations, and development. In J. S. Jordan (Ed.), *Systems theories and a priori aspects of perception* (pp. 47–74). Elsevier Science.

Lee, S. H. (2021). Effects of retailers' corporate social responsibility on retailer equity and consumer usage intention. *Sustainability, 13*(6), 3080. https://doi.org/10.3390/su13063080

Lee, E.-J., Choi, H., Han, J., Kim, D. H., Ko, E., & Kim, K. H. (2020). How to "nudge" your consumers toward sustainable fashion consumption: An fMRI investigation. *Journal of Business Research, 117*, 642–651. https://doi.org/10.1016/j.jbusres.2019.09.050

Lehner, M. (2014). Translating sustainability: The role of the retail store. *International Journal of Retail & Distribution Management, 43*(45), 386–402. https://doi.org/10.1108/IJRDM-02-2014-0013

Li, J., & Leonas, K. K. (2021). The impact of communication on consumer knowledge of environmentally sustainable apparel. *Journal of Fashion Marketing and Management, 26*(4), 622–639. https://doi.org/10.1108/JFMM-02-2021-0034

Lincoln, Y. S., & Guba, E. G. (1985). *Naturalistic Inquiry*. Sage.

Littlejohn, S. W., & Foss, K. A. (2009). *Encyclopedia of communication theory*. Sage.

Ma, Y. J., Gam, H. J., & Banning, J. (2017). Perceived ease of use and usefulness of sustainability labels on apparel products: Application of the technology acceptance model. *Fashion and Textiles, 4*(3), 1–20. https://doi.org/10.1186/s40691-017-0093-1

McNeill, L., & Moore, R. (2015). Sustainable fashion consumption and the fast fashion conundrum: Fashionable consumers and attitudes to sustainability in clothing choice. *International Journal of Consumer Studies, 39*(3), 212–222. https://doi.org/10.1111/ijcs.12169

Meadows, D. H. (1999). *Leverage points: Places to intervene in a system*. The Sustainability Institute.

Mele, C., Pels, J., & Polese, F. (2010). A brief review of systems theories and their managerial applications. *Service Science, 2*(1–2), 126–135.

Modi, D., & Zhao, L. (2021). Social media analysis of consumer opinion on apparel supply chain transparency. *Journal of Fashion Marketing and Management, 25*(3), 465–481. https://doi.org/10.1108/JFMM-09-2019-0220

Mohr, L. A., Webb, D. J., & Harris, K. E. (2001). Do consumers expect companies to be socially responsible? The impact of corporate social responsibility on buying behavior. *Journal of Consumer Affairs, 35*(1), 45–72. https://doi.org/10.1111/j.1745-6606.2001.tb00102.x

Moore, S. B., & Ausley, L. W. (2004). Systems-thinking and green chemistry in the textile industry: Concepts, technologies and benefits. *Journal of Cleaner Production, 12*, 585–601.

Morgan, L. R., & Birtwistle, G. (2009). An investigation of young fashion consumers' disposal habits. *International Journal of Consumer Studies, 33*(2), 190–198. https://doi.org/10.1111/j.1470-6431.2009.00756.x

Niesenbaum, R. (2020). *Sustainable solutions: Problem solving for current and future generations*. Oxford University Press.

Oeko-Tex Made in Green. (n.d.). https://www.oeko-tex.com/

Oeko-Tex Standard 100. (n.d.). https://www.oeko-tex.com/en/our-standards/oeko-tex-standard-100

Oh, K., & Abraham, L. (2015). Effect of knowledge on decision making in the context of organic cotton clothing. *International Journal of Consumer Studies, 40*(1), 66–74. https://doi.org/10.1111/ijcs.12214

Okur, N., & Saricam, C. (2018). The impact of knowledge on consumer behaviour towards sustainable apparel consumption. In S. Muthu (Ed.), *Consumer behaviour and sustainable fashion consumption* (pp. 69–96). Springer. https://doi.org/10.1007/978-981-13-1265-6_3

Orr, D. (2010). *Hope is an imperative: The essential David Orr*. Island Press.

OTEXA. (n.d.). *U.S. labeling requirements or textile, apparel, footwear and travel goods*. https://otexa.trade.gov/us_labeling.htm

Radhakrishnan, S. (2014). The sustainable apparel coalition and the Higg Index. In S. S. Muthu (Ed.), *Roadmap to sustainable textiles and clothing* (pp. 23–57). Springer.

Rankin, S., Hunter, C., & Kramer, M. (2020). *Revive to survive: 2021 retail state of the industry report*. https://advisory.kpmg.us/articles/2020/retail-state-of-industry-2020.html
Rathinamoorthy, R. (2020). Clothing disposal and sustainability. In S. Muthu & M. Gardetti (Eds.), *Sustainability in the textile and apparel industries* (pp. 89–120). Springer. https://doi.org/10.1007/978-3-030-38532-3_6
Ross, M. (Producer), & Morgan, A. (Director). (2015). *The true cost* [Motion picture]. Life is my movie entertainment company.
Rugmark. (n.d.). https://rugmarkindia.de/
Sánchez-González, I., Gil-Saura, I., & Ruiz-Molina, M. E. (2020). Ethically minded consumer behavior, retailers' commitment to sustainable development, and store equity in hypermarkets. *Sustainability, 12*(19), 8041. https://doi.org/10.3390/su12198041
Sides, R., & Skelly, L. (2021). *2021 retail industry outlook.* https://www2.deloitte.com/content/dam/Deloitte/us/Documents/consumer-business/us-2021-retail-industry-outlook.pdf. Accessed 13 August 2021.
Sinha, P., Muthu, S. S., & Dissanayake, G. (2016). Systems requirements for remanufactured fashion as an industry. In S. S. Muthu (Ed.), *Remanufactured fashion* (pp. 45–71). Springer.
Solomon, M. R. (2017). *Consumer behavior: Buying, having, and being*. Pearson.
Spiggle, S. (1994). Analysis and interpretation of qualitative data in consumer research. *Journal of Consumer Research, 21*(3), 491–503. https://doi.org/10.1086/209413
Stanszus, L., & Iran, S. (2015). Sustainable fashion. In S. Idowu, N. Capaldi, M. Fifka, L. Zu, & R. Schmidpeter (Eds.), *Dictionary of corporate social responsibility: CSR, sustainability, ethics and governance* (pp. 514–515). Springer.
Sustainable Apparel Coalition. (2019). *Empowering consumers through transparency: Report on global consumer research.* http://apparelcoalition.org/wp-content/uploads/2019/07/Research-Report-Empowering-Consumers-through-Transparency.pdf
Turker, D., & Altuntas, C. (2014). Sustainable supply chain management in the fast fashion industry: An analysis of corporate reports. *European Management Journal, 32*(5), 837–849. https://doi.org/10.1016/j.emj.2014.02.001
United Nations. (1987). *Report of the world commission on environment and development: Our common future.* (Annex to report no. A/42/427). http://www.un-documents.net/wced-ocf.htm
United States Environmental Protection Agency. (n.d.). *Facts and figures about materials, waste and recycling.* https://www.epa.gov/facts-and-figures-about-materials-waste-and-recycling/textiles-material-specific-data
Vadakkepatt, G. G., Winterich, K. P., Mittal, V., Zinn, W., Beitelspacher, L., Aloysius, J., Ginger, J., & Reilman, J. (2021). Sustainable retail. *Journal of Retailing, 97*, 62–80. https://doi.org/10.1016/j.jretai.2020.10.008
Van der Merwe, D., Bosman, M., Ellis, S., Van der Colff, N., & Warnock, M. (2014). Consumers' knowledge of textile label information: An exploratory investigation. *International Journal of Consumer Studies, 38*, 18–24. https://doi.org/10.1111/ijcs.12053
Vehmas, K., Raudaskoski, A., Heikkila, P., Harlin, A., & Mensonen, A. (2018). Consumer attitudes and communication in circular fashion. *Journal of Fashion Marketing and Management, 22*(3), 286–300. https://doi.org/10.1108/JFMM-08-2017-0079
Von Bertalanffy, L. (1972). The history and status of general systems theory. *Academy of Management, 15*(4), 407–426.
Wang, C.-T. (2020). *Welcome to the great retail reinvention.* https://www.cushmanwakefield.com/en/united-states/insights/us-articles/2020-02-dal-retail-reinvention
Webster, F., Jr. (1975). Determining the characteristics of the socially conscious consumer. *Journal of Consumer Research, 2*, 188–196. http://www.jstor.org/stable/2489054
Williams, A., & Hodges, N. (2022a). Adolescent Generation Z and sustainable and responsible fashion consumption: Exploring the value-action gap. *Young Consumers, 23*(4), 651–666. https://doi.org/10.1108/YC-11-2021-1419

Williams, A., & Hodges, N. (2022b). Signaling sustainability: Exploring consumer perspectives on communicating apparel sustainability information. *Journal of Sustainable Marketing, 3*(1), 26–39. https://doi.org/10.51300/jsm-2022-49

Williams, A., Hodges, N. J., & Watchravesringkan, K. (2024). Testing the effects of an apparel sustainability index on consumers' brand attitudes and evaluations: An application of signaling theory. *International Textile and Apparel Association Annual Conference Proceedings, 80*(1), 10.31274/itaa.17528.

Yan, R.-N., Ogle, J. P., & Hyllegard, K. H. (2010). The impact of message appeal and message source on Gen Y consumers' attitudes and purchase intentions toward American Apparel. *Journal of Marketing Communications, 16*(4), 203–224. https://doi.org/10.1080/13527260902863221

Yan, R., Hyllegard, K., & Blaesi, L. (2012). Marketing eco-fashion: The influence of brand name and message explicitness. *Journal of Marketing Communications, 18*(2), 151–168. https://doi.org/10.1080/13527266.2010.490420

Yudina, A. (2017). *The Higg index: A way to increase sustainable consumer behavior?* [Master's thesis, University of Colorado Boulder]. https://scholar.colorado.edu/concern/undergraduate_honors_theses/8049g5435.

Zheng, Y., & Chi, T. (2015). Factors influencing purchase intention towards environmentally friendly apparel: An empirical study of US consumers. *International Journal of Fashion Design, Technology and Education, 8*(2), 68–77. https://doi.org/10.1080/17543266.2014.990059

Chapter 3
Designing Sustainable Fashion for Circular Behavior

Muhammad Babar Ramzan, Jawad Naeem, and Hajra Mustafa

3.1 Introduction

The fashion and textile industry are normally considered as the primary source of pollution on the earth. This is due to the fact that the fashion industry is responsible for generating approximately 92 million tons of waste each year with an expenditure of 79 trillion liters of waste (Peters et al., 2020). Fast fashion has acquired prosperity in the international market and has become a global trend (Jung & Jin, 2016). It facilitates swift response time to marketing strategy, less expensive, and demand-driven fashion (Anguelov, 2015). The production of fast fashion is dependent on the demands of the market as fast as possible (Caro & Martínez-de-Albéniz, 2015). The values inculcated in the business model of fast fashion enhance buying and consumption rates spontaneously (Johansson, 2010). This causes the problem of overconsumption and landfill waste in huge amounts. The waste consists of pre-consumption and post-consumption waste.

Fast fashion proposes economic price and short term-life trendy products within weeks and severely depends on the desire to purchase (Anguelov, 2015). As a result, the consumption pattern of consumers has altered and the current era's consumer has a swift approach to purchasing fashionable trendy clothes than ever before (Joy et al., 2012). With the change in trends of fashion, the necessity to spend on clothing also enhances inevitably. At present, a lot of attention is given to the problem of overconsumption of clothing (Mishra et al., 2020). The main reason behind this

M. B. Ramzan (✉) · J. Naeem
School of Engineering and Technology, National Textile University, Faisalabad, Pakistan
e-mail: drbabar@ntu.edu.pk; jawad@ntu.edu.pk

H. Mustafa
Institute of Art and Design, Government College University, Faisalabad, Pakistan

S. S. Muthu (ed.), *Sustainable Apparel Retailing*, Sustainable Textiles: Production, Processing, Manufacturing & Chemistry,
https://doi.org/10.1007/978-3-031-83875-0_3

issue is the accomplishment of fast fashion. Both the consumers and clothing companies are accountable for consumption manners. Now customers have an obsession for consumption. It is one of the most important challenges in shifting from fast fashion to sustainable fashion (Peters et al., 2020).

The combination of fast fashion with irregular dumping methodology is increasing fear among environmentalists as our landfills are stacked with billions of kilograms of post-consumer textile waste (Sandin & Peters, 2018). Despite the terrifying level of landfill wastage, much research has not been conducted in the post-consumption stage as compared to the pre-consumption stages (Ki et al., 2020).

Kate Fletcher mentioned that a reduction in the value of presented products is one of the major reasons for disposal of garments. This might be due to the production of short-life garments, less costly, and behavior of the present customer and this blending is the outcome of fast fashion (Grose & Fletcher, 2012). By inculcating value addition in these products, the problems of reduced value of these products can be handled effectively. The product must be developed desirably along with the removal of waste so that the product will not stacked in a landfill after a short period of time (Saha & Lehew, 2018). K Saha (Saha & Lehew, 2018) mentioned the durability of garments in the form of physical properties might not be an appropriate answer for overconsumption of fast fashion as lack of durability is not the major reason for enhancement of post-consumption wastages. Therefore, it was suggested that apart from physical durability, there must be emotive durability for value addition. The customer must play a huge part in forcing the fashion industry to shift toward sustainable fashion and switch their behavior from unsustainable consumption behavior to conscious behavior (Claudio, 2007). By educating and creating awareness among the customers might not be sufficient to alter the behavior considerably, but the designers and makers can impact modification in consumption behavior of customer through the value addition concept (Morlet et al., 2017). This study reviews different sustainable strategies for enhancing the life cycle of garments and reduction in pre-consumer waste and post-consumer waste. For this purpose, almost 100 articles were studied, and 72 articles were selected. After the section of the introduction, slow fashion was discussed. Afterward, circular fashion was discussed, involving designing for circularity, designing for disassembly, and designing for sustainable fashion. Later, previous research related to sustainable strategies involving pre-consumer and post-consumer wastes was discussed.

3.1.1 Slow Fashion

The development of slow fashion is framed on contrasting side of spectrum from the fast fashion generation model. It materialized as an opposing part of fast fashion. Kate Fletcher (Jung & Jin, 2014) discussed that slow fashion movement was originated from the "slow food" movement that facilitates sensible food production and its utilization. Garment manufacturing companies and brands that backed sustainability inculcates production which is more ethical in terms of sourcing of materials, slowing of production process along with consumption and break down of barriers between stakeholders and organizations and empowerment of labor. Slow fashion inspires consumers to give more weightage to quality as compared to quantity. Moreover, it advocates ethical practices inside and outside of organizations (Fletcher, 2010). Slow fashion contests traditional ways of industry and advocates variation in thinking, throughput time of goods and frame of organization. It stresses fundamental change which is associated to more transformational and universal approaches which encounters mainstream consumer culture and tasks outside the system. Slow fashion emphasizes the problems of overconsumption and leads innovative business model and request for individualistic involvement and awareness to encourage change in social behavior.

Fletcher stated that it is feasible to bring in environmentally friendly and social change for integration of sustainable principles in design and manufacturing procedures. All this can be probable with the utilization of sustainable design strategies, i.e., alignment of fashion with sustainability despite surrendering fashion (Fletcher, 2007). A lot of attempts have been made in the textile and apparel industry for sustainable design with a major focus on materials and textiles. Therefore, the utility of virgin material and the flow of textile are utilized as gauge of influence on the environment. The proposed answer is the use of less virgin material, i.e., blend of virgin and recycled materials in manufacturing process (Grose & Fletcher, 2012). However, apart from usage of organic and recycled materials, there are lot of possibilities beyond that and these are known as sustainable design strategies (Rissanen, 2008).

Gwilt reported (Alison Gwilt, 2011) the current known sustainable design strategies as compared to traditional design procedure. Figure 3.1 demonstrates approach of sustainable design through the utility of design strategies which impact the production process from sketching design of garment to end product or ahead of that. It inculcates strategies like design for disassembly, design for social well-being, design for the end-of-life strategies, design for waste minimization, design for user-participation, design for slower consumption, and design for product service systems.

These strategies facilitate business models that are not like traditional business models. However, it leads to viable encounters as it switches from a linear business model to a circular business model.

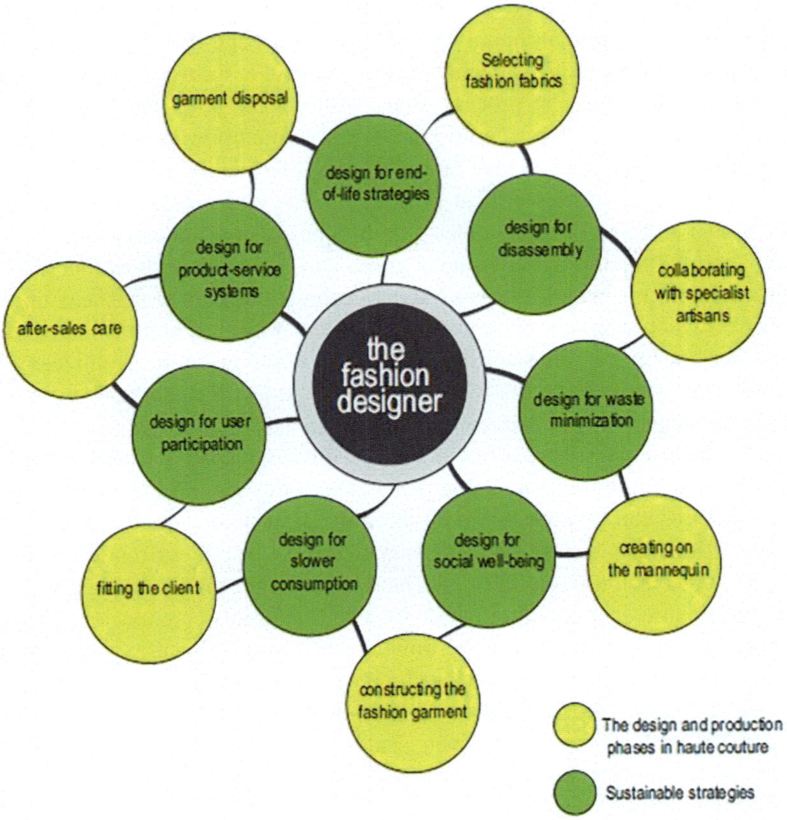

Fig. 3.1 Sustainable design strategies (Alison Gwilt, 2011)

3.2 Circular Fashion

A new concept called circular fashion (Ki et al., 2020) means designing fashion products by looking at factors like the longevity of products, ethical practices, non-hazardous recyclable products, efficiency in the utility of resources, and biodegradation. Its main aim is to keep the process in circulation. Product life extension is related to maintenance, the ability of repairability, upgradation, compatibility, disassembly, and reassembly (Linton & Jayaraman, 2005). This approach is about widening the utility tenure of product through accessibility of service circle. Champman (2005) discussed that longevity of product life is assisted by characteristics that evolve association of entrust and authenticity between consumer and merchandize, i.e., designating emotional endurance (Chapman, 2005) and physical robustness (Moss, 1985).

3.2.1 Designing for Circularity

To counter the problem of fast fashion trends, design for circularity has become a global aim of the fashion business. The circular business model maintains the utilization of raw materials in the loop. It permits the possibilities to reuse materials and goods, where there is feasibility of utilization of renewable resources. Chong et al. (Ki et al., 2020) mentioned that research in the field of circular fashion appeared in 2017 with a total number of 13 circular fashion published articles. The concept of circular fashion was introduced in 1990s. However, research in academic institutions was not done at huge level.

Bocken et al. (2016) investigated the literature on circular design and business model approaches and stated that both these strategies brought forward approaches for closed-loop design and business prototypes which are the opposite of open-loop enterprise and business models. Bocken et al. (2016) segregated business enterprise approaches corresponding to the procedure through which sources passes into and out of the system. Braungart et al. (2008) distinguished "cradle to grave" and "cradle to cradle" (McDonough & Braungart, 2002) flows of the substrate, which created the differentiation between pattern of source flow in process of product manufacturing. Bocken et al. (Bocken et al., 2016) presented three different approaches toward the circulation of sources, i.e., closing source loop approaches, limiting the source loop and reducing the loop approaches as demonstrated in Fig. 3.2.

Reducing flow of resources is related to sustained utility of goods and its reusage through extension tactics employed to the products (Bocken et al., 2016). Tightening the loops is related to declining the rate of assets utility connected to the product development process. However, closing of loops facilitates the reusage of substrates through the process of recycling (Bocken et al., 2016). Moreover, it was recommended that reducing and limiting the loops might be a similar, therefore development on two basic tactics for resource cycling, i.e., closed and slowing down the resource loop was investigated from the literature to revive approaches for circular economy. Table 3.1 demonstrated two specific tactics which are under shadow of circular design product.

3.2.2 Design for Disassembly

Design for disassembly (DFD) originated from design for assembly (DFA) and design for manufacture (DFM) and these theories were developed in 1970 (Bogue & Bogue, 2007). DFD enhances the recycling, reusability, and reproductivity of substrates and elements. This theory intersects and participates in the development of biological and technical pattern (Bocken et al., 2016). DFD and DFR theories exhibit the disassembling of products and their components or easy disintegration and integration of materials and products over again (Bakker et al., 2014). This approach assisted in incrementing the rate of reusability of substrates and its elements (Short, 2014) and confirms the importance of separating materials under various cycles, i.e., biological or technical. Therefore, it was kept in mind that DFD

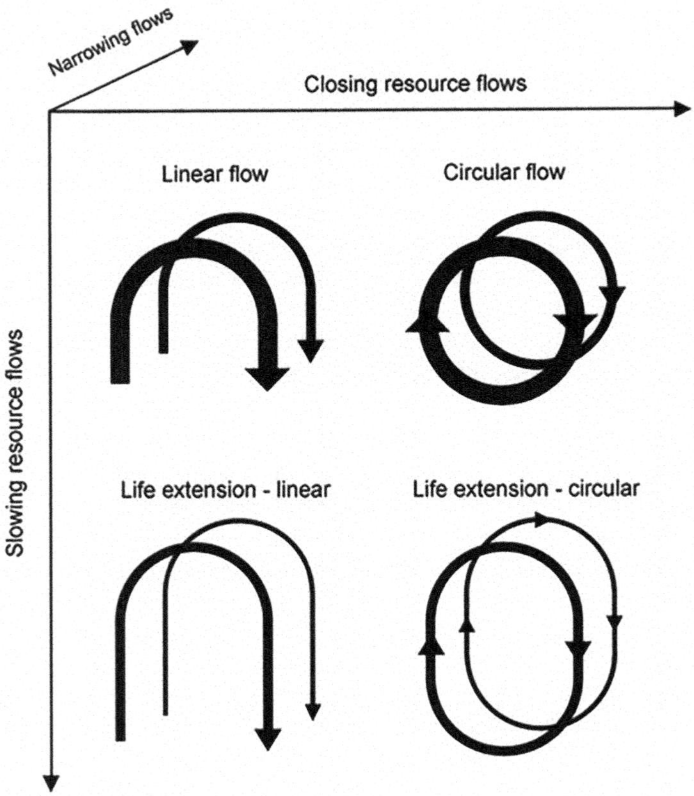

Fig. 3.2 Categorization of circular and linear approaches to reduce resource use (Bocken et al., 2016)

Table 3.1 Product design strategies (Bocken et al., 2016)

Circular product design strategies	
Design strategies to slow loops	**Design strategies to close loops**
Designing long-life products Design for attachment and trust Design for reliability and durability	Design for a technological cycle Design for disassembly and reassembly Design for a biological cycle
Design for product life extension Design for ease of maintenance and repair Design for upgradability and adaptability Design for standardization and compatibility Design for dis- and reassembly	

manufactured goods are developed for disassembly and designers are major factors in the incorporation of these kinds of design elements into products. DFD approach inculcates the closing resource loop approaches which can suit both approaches as mentioned in Table 3.1. To make it feasible, the product is developed in a way that its components can be effortlessly assembled, disassembled and reassembled

(Bocken et al., 2016). R. Bogue (Bogue & Bogue, 2007) inferred that a decline in post-consumer wastages is the major factor for facilitating DFD. At present, the existing procedures of production are considered unsustainable, and the examination of the life cycle of a product is the focus of all kind of industries. Implementation of DFD is used in manufacturing industrial products, along with architecture and several other fields. Nonetheless, this approach is still not attracting designer and researchers in the domain of fashion designing.

3.2.3 Designing Sustainable Fashion

3.2.3.1 Sustainable Design

Designing sustainability is contemplated as a significant factor in encouraging ethical values for the cause of sustainability, as the cost of environment of product is evaluated by its design (Design Council, 2013). This showed the importance of role of design in acquiring sustainability. The word design can be referred to both processer and the end product (Lawson, 2005). Despite all these facts of embodiment of sustainable values in processes and products, the focus is more on the process side of product. Jonathan Chapman (2015a) mentioned in "emotionally durable design" that design has the capability of changing our perspectives of the world, supported association with the people and alter our views and influence our attitude (Brass & Bowden, 2012). In conclusion, requirement of design is essential and beneficial procedure for novelty and invention (Chapman & Gant, 2015), it was proposed by many researchers to employ it as capable device in advancing toward sustainability.

Walker discussed the initiation of novel concepts and making them triumphant by emphasizing going further beyond conventional boundaries of design. Rather than pushing sustainable ideas into traditional procedures, the designers should focus on manufacturing products which are correspond to sustainability (Walker, 2006). An excellent design must have enduring value inculcating to it. On the other hand, it must not be unremembered that most characteristics that make the product suitable and significant might alter after a certain period and in the framework of sustainability mean a decrease in value. Walker (Walker, 2006) reported that the attractive appealing and functional capability of manufactured goods fades with time undoubtedly. Due to this reason, designers must put a lot of effort into the value addition of products to increment product life. Sustainable design must be developed in such a way that both material and product retaining considerations must be kept in mind. Longevity of manufactured goods must amend both emotional and physical durability to resolve the problem of overconsumption (Goldsworthy, 2017). This precisely reduces the rate of propagation, i.e., slows down the rate of production and consumption. With the increase in the circular economy, designers and researchers are constantly progressing and studying circular methodologies and aids for the development of products. Most of them suggested enhancement of the life cycle of the product along with emotional lifetime. However, many researchers are proposing design for loyalty. Most of the business theories and models and aids were proposed to support the designers and manufacturers (Medkova & Fifield, 2017; Haines-gadd et al., 2018). Most of the designers suggested design for attachment and belief as basic principle for circular design. However, a lot of research is required in this field. Bakker and Hollander suggested development of products while bearing in mind both emotional and physical durability for longer life cycle of product (Bakker & Schuit, 2017). These two features of durability have demonstrated the ability to tackle the problem of overconsumption which is obstruction to sustainable approach. Saha et al. (Saha & Lehew, 2018) suggested and emphasized the importance to highlight both features of durability for considerable modification in the attitude of customer.

$$\text{Functional durability} + \text{Emotional durability} = \text{Sustainable Apparel Consumption}$$

The issue of overconsumption existed not due to frangible manufactured goods but due to the deficit of accessibility of emotional ingredients in the manufactured goods. Flecther (Grose & Fletcher, 2012) reported deficit of meaning and compassion in fashionable products along with low cost and easiness of buying improves the portray of abandoning the product before they were physically worn away. Sometimes the disposal of a product is proof of unsuccessful association, not the impoverished quality as it is not successful in establishing an association with the consumer (Saha & Lehew, 2018). The product becomes visible as a liability rather than an asset occurred when manufactured goods is unsuccessful to establish between consumer and itself (Grose & Fletcher, 2012).

Emotional durability inculcates strategies like Design for X which includes design for disassembly, design for recycling, and design for reuse, etc. The main aim is to establish successful association between a product and consumer. It concentrates on developing a stronger connection between material things and the users and seeks decline in rate of consumption and wastages (Chapman, 2015a). Chapman (2015a) reported wastages as a display of no connection and unsuccessful association. By developing emotionally durable products, manufacturing organizations required not to produce goods as swift speed, but it can focus on establishing improved products rather than providing services to assist in preserving the products. In fashion industry, the manufactured goods are disposed because of two reasons. First, the change in trends and second is unsuccessfulness of encouraging connection between consumers and products.

The most important property of stable design is related to permit products to age alongside the consumers and to establish association between the consumer and product. Chapman (2015b) stated in existential ideas that assists in incrementing the passionate association between product and the user. Haines et al. (2018) collected and mentioned strategies by a lot of researchers for association of manufactured goods as mentioned in Table 3.2. By inclusion of these properties in the product might influence in longevity of product life. One approach is to allow people to utilize products for extended period of time by developing products that can be modified with time and upgraded with age (Chapman, 2015a). It can be performed by adding properties like transformation and adaptation capability to goods and services.

Table 3.2 Key strategies for product attachment

Strategies	Researchers
Memories and (longevity) (nostalgia)	Mugge (2007), Schifferstein and Zwartkruis-Pelgrim (2008), Maclachlan (2011), Page (2014)
Pleasure	Mugge (2007), Schifferstein and Zwartkruis-Pelgrim (2008), Maclachlan (2011), Page (2014)
Enjoyment	Schifferstein and Zwartkruis-Pelgrim (2008)
Self-expression or support-self identity	Mugge (2007), Schifferstein and Zwartkruis-Pelgrim (2008), Maclachlan (2011)
Usability	Page (2014)
Sensory design	Maclachlan (2011), Ludden
Superior appearance	Mugge et al. (2010)
Utility and reliability	Schifferstein and Zwartkruis-Pelgrim (2008)
Product personality	Mugge (2007), Maclachlan (2011)
Group affiliation	Mugge (2007)

3.2.3.2 Sustainable Design Strategies and Waste Management

The purpose of sustainable design is to find novel ways of developing waste and generating ethical products. The usual way in the industry is to minimize the harmful impact of production and design. McDonough and Braungart (2002) employed terminology like reduce, sustain, halt, avoid and limit, and refer to investigate approaches that are less likely to draw criticism than traditional methods. These approaches focus on declining the faults of the present system rather than modifying the entire system. However, according to Lawson (2005), there is a demand to probe the frequent procedures in the design field.

Sustainable procedure can be executed both in the pre-consumer and post-consumer phase to acquire reduction in waste. Usual waste management approaches are: reduce, reuse, and recycle. These approaches are primarily named as 3R (Fletcher, 2008), and their main aim is to enhance the life cycle of a product. These approaches are a part of a larger model distinguished as waste management hierarchy as presented in Fig. 3.3, presented by Timo Rissanen and based it on the theory of Franke, Hindle, and White (Gaugele, 2008). According to Timo, waste inhibition is the most preferred choice while the least favorite is waste disposal.

Most approaches have been developed for the progress of sustainable design strategies with differing success and failure ratios. Rebecca Earley (Earley et al., 2010) mentioned ten widely known strategies for design field distinguished by textiles environment design project. These are design to substitute the necessity to consume, design for ethical productions, design to decline the energy and water utility, design to reduce waste, design for recycling, design to decrease chemical impacts, design to disperse and evolve systems and services, design that discovers clean and improved technologies, design involvement, and design that looks at reproductions from nature and history.

Sustainability demands such attempts that inculcate developing most of the products at their ending stage of life to deliver as much resources as they can into the

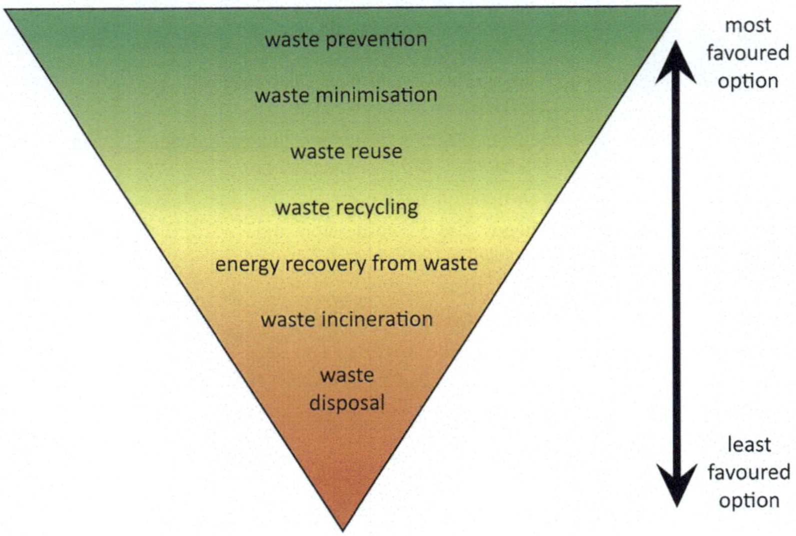

Fig. 3.3 Waste management hierarchy (Gaugele, 2008)

value chain, by concentrating on the post-consumption phase. This concept is promoted by approaches that are centered on steps of product life cycle. These approaches are known as "Design for X" (Dfx), inculcating approaches such as design for disassembly (DFD), reuse, and recycling. Dfx strategies are most commonly adopted approaches and its aim is to enhance value to the already used textile and thus give them a new life (Bocken et al., 2016). These approaches are in the center and slightly above center in the hierarchy framework mentioned by Rissanen in Fig. 3.3.

3.2.3.3 Zero Waste Fashion

Zero waste fashion sponsors above-mentioned design to reduce waste (Earley et al., 2010; Rissanen, 1994). Approximately, 85% of the fabric spread in a clothing is being used in the traditional methodologies for garment manufacturing. Due to this reason, 15% of wastage is produced (Carrico & Kim, 2014). Waste reduction at the design stage like in zero waste fashion design ZWFD decreases the production waste to a minimum level. Zero waste fashion is proposed to create garments with minimum or zero textile waste, this method can be employed to the design stage along with cutting of fabric and pattern, and production stage for decrease in waste and intention of eradication. Timo Rissanen asserted that the most beneficial approach is to eradicate waste in the design stage (Gaugele, 2008).

Holly McQuillan reported ZWFD as favorite choice in waste management approach as its focus is reduction in waste during production (Derwin, 2016). There are several methods created by ZWFD pioneers on garment manufacturing. Some of them are geo-cut and drape and cut, subtraction cutting, jigsaw puzzle, cut, and sew, A-POC, fully fashioned. Zero waste fashion focusses on both pre-consumer waste and post-consumer waste (Saeidi & Wimberley, 2017). Pre-consumption waste inculcates design cutoffs, inventory losses, end roll waste, and damaged items. Post-consumption waste comprises of textiles that have passed the phase of consumer usage. From the previous literature, it was noted that pre-consumer phase has been investigated several times as compared to post-consumer stage despite the problem of landfill waste.

Most of the issues in the current fashion industry like generation of waste can be prevented by integration of design and production stages. This situation does not suggest transferring of roles rather mentioned increased association between the fashion designer and the pattern cutter. McLennan (2004) have the same opinion as of Loschek (2009) and suggested that enhancing cooperation (communication) between different persons like designers, pattern cutters, and managers could possibly assist in design procedure for an efficient design. This idea of partnership is significantly encouraged by Zero Waste Fashion Design (ZWFD). The ZWD design procedure is distinct from the traditional design process as seen from Fig. 3.4 (Rissanen, 2013). In the conventional design procedure, the pattern cutter is accountable for effective utility of fabric to evade wastage. Whereas in the ZWD process, it is the responsibility of the fashion designer to acknowledge this problem while designing/sketching in the first place (Rissanen, 2008). Rissanen (1994) discussed that fashion designer responsibilities should also inculcate effective fabric usage procedures. Rissanen and Cooklin stated that textile fabric is the utmost expensive raw material and concoct approximately 50–70% of the total cost of the garment (Rissanen, 1994). Furthermore, labor cost is the most substantial source of garment cost. Hence, there is no option to waste such costly raw material and researchers along with some designers proposed that waste management approached can be investigated to reduce hazardous impacts of industry.

Holly McQuillan (2020) also suggested that these kinds of cooperation takes into account the waste inhibition attribute and suggested a bodywork as a replacement to present linear model as shown in Fig. 3.5. In the prevailing garment manufacturing system, there is a segregation line between the design process and pattern cutting process, as it is apparent that contemplation of pattern cutting at the primary stage

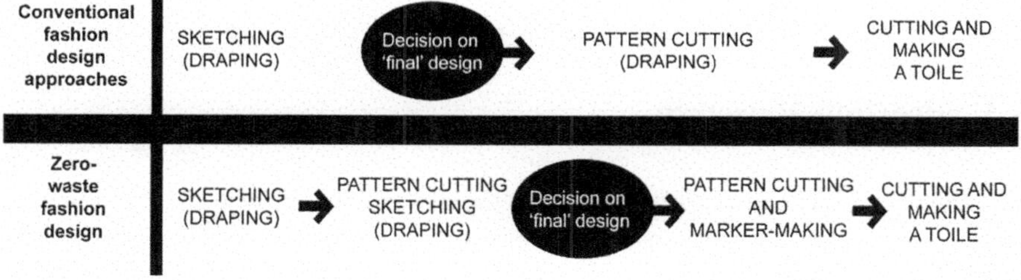

Fig. 3.4 Traditional/Conventional design process vs ZWD process (Rissanen, 2013)

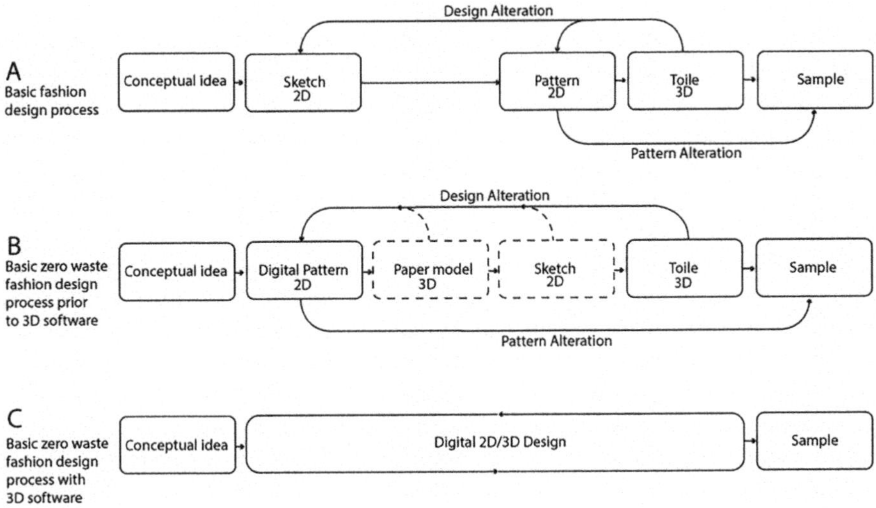

Fig. 3.5 Diagram showing comparison between three fashion design processes, (**a**) developed by (Rissanen, 2005) categorization, (**b**) a basic zero waste design, and (**c**) a basic zero waste design process using 3D tools (McQuillan, 2020)

of designing might disturb or may cause loss of design creativity. McQuillan and Rissanen (McQuillan, 2019) reported in favor of zero waste design techniques and affirmed that it focuses to eliminate this hedge as it vows for hybrid acts by the fashion designer and pattern cutter. McQuillan advocated that employing strategies like zero waste fashion design and its execution through 3D software can pace up the manufacturing procedures along with reduction in waste.

James et al. (2016) also conducted research to explore the level of influence ZWFD can have on the traditional designing process of garments. The second goal was to investigate the connection between the fashion designer and the pattern cutter. To examine both objectives of this study, the first practice-based experiments were conducted that implied ZWPC. In the second stage, the production process was observed through interviews with pattern cutters and fashion designers. It was determined that ZWFD encourages partnership between the designer and the pattern cutter, while improving the iterative process of designing, by adjusting the course of production.

3.3 Previous Research

Niinimaki et al. (Peters et al., 2020) a review paper, mentioned all the essential items in the value chain of fashion commerce where it was yielding waste. The textile and fashion industry has a massive supply chain, starting from making fiber and concluding at production and retailing. Waste creation embarks at the very primary level of production and utility of water, substrate, chemical, and energy usage. Due to this fact, every stage of productivity has its influence on the surrounding climate.This study highlights the problem of worldwide textile wastage and the prevailing method of expenditure is resulting in tons of landfill waste every year. It spotlights typically on water usage, chemical pollution, and carbon dioxide emissions. An observation was brought up in terms of incrementing consumption attempts to progress sustainability. The outcome of statistical data revealed increase in global apparel consumption to around 62 million tons and projected to reach 102 million tons by 2030, and this was partially due to fast fashion business model. In this review paper, recommendations were mentioned to reduce the hazardous influence of industry on the climate by declining fabrication of garments and their wastages along with extension of garment utility and life cycle.

Business models left considerable influences on the surrounding climate environment in which they are followed. A right knowledge of a business model mandated an effective study further than the connection between the producer and consumer, it should take the complete model into consideration. For instance, the model depicted in "Sustainable Consumption and Production (SCP); a handbook for policymakers by United Nations Environment Program" (UNEP, n.d.) can be seen in Fig. 3.6 emphasized on life cycle thinking which entails how manufacturers must focus their attention on their production process in order to inculcate

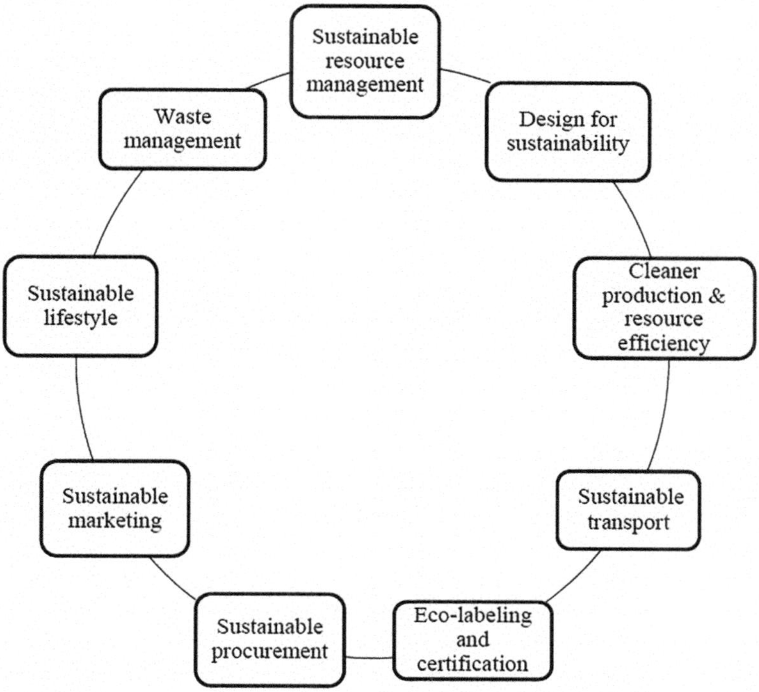

Fig. 3.6 Sustainable consumption and production (SCP) cycle

principles like cradle to cradle model (C2C). Apart from C2C models, business models did not consider the issue of what will occur to the product at the end of its life. SCP cycle has nine fundamental principles that highlighted circular way of reasoning and configure that consumption is as pivotal as manufacturing. As a result, elimination of one will make all the efforts unsuccessful. Evans et al. (Fossen, 2017) also emphasized on how compulsory it is for businesses to introduce sustainable standards and practices throughout their entire business models. They put prominence on the insertion of environmental sustainability as one of the central drivers in the business models.

Life cycle theory models hold distinction with the traditional linear business model of manufacturing of take-make-utilize-dispose. Kerr and Landry (Eder-Hansen et al., 2017), conducted a study in 2017 by Global Fashion Agenda (GFA) and the Boston Consulting Group (BCG) for the fashion summit in Copenhagen, it was determined that 78 companies still use conventional linear models in fashion out of 91. However, only 13 companies incorporated sustainable value into their operational modes. This linear business model was shaded by fast fashion business model. Fast fashion has a disadvantage in terms of environmental sustainability. It facilitated economic gain over the environmental and social components, causing harmful impacts.

To restrain the issue of industrial solid textile waste (fabric waste), McQuillan (2019) proposed that hybrid zero waste design approaches were needed for a holistic method toward production waste, as the prevailing fashion production procedures were declared unsustainable. For example, the crossing between fashion and textile design to decrease or eradicate waste completely at the design stage and discovered novel terms. ZWFD schemes can be executed in the designing level for waste exclusion. James et al. (2016) also conducted a research to inquire the level of impact ZWFD can have on the traditional designing procedures of garments. The second aim was to explore the association of the designer and the pattern cutter share. To address both aims of this study, initial practice-based experimentations were conducted that encompassed ZWPC. In the second stage, the production process was examined through interviews with pattern cutters and designers. It was determined that ZWFD does not alter the traditional linear design system but advocated partnership between the designer and the pattern cutter. Decline of expenditure of material during manufacturing process is the center of Zero Waste Fashion and this policy was also flourishing to elaborate efficient schemes for increased consumer engagement (Derwin, 2016).

Derwin and Holly McQuillan in another research conducted a study founded on an interdisciplinary venture and represented a stratagem for "User Modifiable Zero Waste Fashion UMZWF." In that venture, she utilized textile print as an embedded coding system for the consumer to utilize as a guiding code during garment manufacturing. The investigation was accompanied through in-house testing, which was monitored by workshops for volunteer contributors. It was determined that zero waste matrix delivered the consumer with uninterrupted adjustment opportunities, UMZWF incremented the garment's life cycle and entrenched way-showing system through textile print can be proved supportive if the traditions of fashion industry like knowledge possession are shattered. For further studies, efforts must be made on the combined exclusion or reduction of pre-consumer and post-consumer waste through the eyes of zero waste patterns for elongated life cycle of garments. Zero waste fashion inventors have been trying to foster tactics for waste reduction in the design level, and so numerous sustainable design methodologies are available today that are declared authentic. Carrico and Kim (2014) offered another methodology for ZWPC as a fifth design practice "nominal cut" other than the four design procedures previously shown by the primal zero waste fashion design. Holly McQuillan in her writing "Shaping Sustainable Fashion" mentioned practice-based research that explored the already available zero waste garment procedures and inferred some results. It was determined that marginal cut can be headed to learners for zero waste procedure as it inculcated less cutting.

By conducting case studies in the fields of research, industry, and education, McQuillan (2020) attempted to address the issue of prototype waste and examined the potential contribution of 3D software such as CLO to bolstering zero waste design in clothing construction. It was not considered practical to fully consider cost, productivity, aesthetics, and operation. In the current linear system of fashion production, it was considered an unrealistic and impractical process.

James et al. (2016) in their study stated that zero waste pattern cutting (ZWPC) technique was investigated for the aim of converting the subsequent process of fashion production. In the traditional fashion production processes, pattern makers are only recognizing the vision of the fashion designer with no involvement in the designing process but implementation only, which certainly causes communication problems afterwards. The purpose of the research was to scrutinize to which extent ZWPC modifies the conventional manufacturing process and to discover the transformed connection between the designer and pattern cutter and to eradicate the possible communication matters between the fashion designer and pattern cutter to strengthen association utilizing ZWPC methodologies. In this study, a few experimentations were supervised to examine the impact level of ZWPC on conventional development of productions, which generated number of hypotheses. Furthermore, the conventional process was further explored through consultations with designers and pattern cutters from luxury brands to premium to mass-market retailers and a small artisan design company, which concluded that interior systems could differ. Outcomes showed that the implementation of Zero waste pattern cutting methods advocated cooperation between the designer and the pattern cutter because of re-creation and iteration often.

In the case of post-consumer waste reduction, novel tactics are evolving. Hvass (2014) conducted a study, concentrated on seeking motives behind enterprises commitment with reuse and recycle methods, accessible opportunities and confrontations. Research was organized in the shape of 12 in-depth interviews with seven different fashion companies followed by study of two of them. It was observed that post-retail accountability is a promising discipline, and it poses several prospects for the fashion industry. Nevertheless, it also requires broader investor participation. The two major strategies recognized were secondhand clothing (SHC) and product take-back policies. For further work, it was investigated that post-retailing can have an prominent effect on textile waste reduction and not sufficient amount of work was done due to the novelty of this field.

Lewis et al. (2017) conducted a project concentrating on the development of an efficient repurposing apparel productivity method for SHC garments for 100% utility of post-consumption waste. Plentiful accessible means of SHC were employed as a raw material for the aim of remodeling of SHC apparel. Four garments were constructed utilizing 11 things of men's SHC employing software CAD to enhance the production of SHC elements by arranging miniature pattern parts inside the bigger pattern pieces and in this way advanced arrangement was obtained using technology. Among used material, out of all the material 36% of the resources were utilized and 64% could not be employed for the creation of new products. Furthermore, small scraps "fiberizer" equipment was designed to meet accomplishment. Investigators proposed that deployment of SHC excludes any necessity for developing new items that exhaust a significant portion of drinkable water and other reserves.

To resolve problems of post-consumer waste, Jin et al. (2011) did research to examine the function of design to disassemble strategy in the creation of men's jackets. As this proposition would lead toward less consumption, increased garment life cycle and reduction, recycle and reuse approaches. A men's jacket from the marketplace was inspected to recognize the barriers to disassembly. Afterwards, a new configuration with characteristics of disassembly was devised and assembled and the assembling time was also documented. Design was concentrated on material assortment, stitch estimation, selection, and design of the jacket. It was determined that reduction in material divergent, precise stitch selection and minimizing marginal details opened a way for a jacket that could easily be undone in one and half minutes. Further research can be done on consumer's behaviors toward design to disassemble methodology in clothing.

Apart from merely emphasizing the fashion industry to take post-retail accountability, it would be beneficial to instruct the wearer about the seriousness of circumstances. Actions like recycling, reusing, and redesigning can be beneficial for deterring waste (Bhatt et al., 2018). Farrant et al. objective was to explore the importance of post-consumer waste decline to reduce textile pollution through several approaches like secondhand clothing (SHC) and reuse, etc. The study was done in the form of a survey questionnaire filled by more than 200 customers in secondhand shops. The survey was managed in Sweden, and Africa was involved by estimation because of more imports of SHC in Africa. The results from the survey indicated that the purchase of 100 secondhand garments could substitute around 60–85 new garments, with 30% decline of waste and 29% reduction in human toxicity in soil. This proved that SHC has a substantial influence on environment and can certainly assist to decline textile carbon footprints. It reduces natural resources utilization, consumption, purchase, and proved less expensive. Researchers proposed that there is a necessity to promote SHC on massive scale with proposed benefits like less costly and environmentally friendly contention.

The philosophy of excessive shopaholism mingled with acute dumping methods is polluting the earth to alarming levels (Laitala & Klepp, 2015). Researchers in previous studies have detected repetition as a key gauge of boredom (Loewenstein & Ubel, 2008). YouGov and Marks and Spencer in a survey of 2229 U.K. customers discovered that 22% of the consumers throw away clothes due to monotony. Additionally, a background of disposable philosophy and ever-evolving trends encouraged wearers to buy more in order to further blend in. Because consumers have limited space to store apparel, they are likely to throw away old items in favor of new ones because styles fluctuate and boredom depends on other factors as well (Hasegawa et al., 2012; Redden & Galak, 2013). Probably, it has conceded in diminishing lifetime of a garment in the wardrobe (Lang et al., 2013).

Kwon (2019) did a study aiming to determine the explanations behind the fast-paced dumping of clothing even though that garment is still in a good physical condition. In this research, the main aim was to examine the chief motive behind boredom and to analyze the level of its effect on retaining, recycling and abandonment of garments. The researchers utilized subsequent hybrid methodology tactic: qualitative phase encompasses thorough interviews to breed hypotheses, and quantitative phase evaluated the hypotheses handling the data gathered from surveys. It was determined that the reason behind boredom mainly incorporates the urge to buy new garments, garment uselessness, and societal acceptance. These outcomes concluded that physical and aesthetic features and situational factors communally affect sentiments of boredom toward garments hanging in the closet. Furthermore, the elements responsible for managing effects on boredom incorporates connection and expedient input, i.e., capital, work, and time. It was determined that wearers particularly exhibited sensitive-laden disposal methodologies and this thing thrived for a communal effort by users, retailers, and officials to separate from these extravagant customs.

Laitala (2014) conducted a study to incorporate experimental research literature on garment dumping in the past to determine general results and research methodology for further research. After studying literature and data assortment, it was known that clothing discarding studies were performed on four major topics: (1) dumping methodologies, (2) motivations behind discarding method selection, (3) reasons for dumping, and (4) gender of consumers tangled in dumping practices. Numerous studies have shown that consumers prefer to donate and leave clothing for others to reuse, such as friends and family. Nevertheless, excessive amounts of clothing are still discarded due to genuine reasons like: transformation in trend, fashion/boredom, wear and tear, stains or other impairments, and occasionally storage/space problems. Some studies revealed that wearers were not taught enough about other recycling theories and alternatives. The review of literature revealed a space about the question of why garment is dumped in the first place. Instead of concentrating on disposal destination, the explanations must be explored that are behind this a of attitude of wearers.

Morgan and Birtwistle (2009) did a study in two stages, concentrating at recognizing how the wearer dumped fashionable clothing and how to improve sustainable intake. For this reason, qualitative–quantitative; triangulation methodology was employed for this exploratory research. The first phase inculcated ten consumer focus groups and six consumer thorough interviews, pursued by the second stage involving an organized survey allotted to young ladies. The focus groups focused on 71 young female consumers to examine their behavior toward buying and product dumping. The six thoroughly designed interviews were taken from ladies aged between 27 and 57 to detect contrasts in behavior with deviation of age. The second stage was governed in Glasgow with a survey form filled in by 225 adult ladies out of which 206 questionnaires were agreeable. It was concluded by the data that focus groups had no detailed idea of how long they would keep a garment in closet. They ceased wearing a typical garment for three major reasons: fashion uselessness, lesser quality, and because garments were bought for one-off event. They were inclined toward keeping expensive garments and swift dumping of less expensive products. It was determined that the users were unaware of the requirements for garment recycling, and they confirmed that fact of lack of knowledge of how and where clothing was dumped. This research about recognizing users' garment dumping attitude emphasized on customer's attitude to be transferred for a noble reason of ecological sustainability and it can be performed through the actions of instructing the user.

Hajra (2021) and Ramzan et al. (2023) conducted a study to examine the potential of Design for Disassembly (DFD) in augmenting the value of clothing for life extension, redesign, and in altering the consumption attitude of wearer. In this aspect, three female subjects (age group 25–35), two garment types (Tunic and Kaftan), and three stitch types (Lock stitch, Hand stitch, and overlock) were chose. A total of 18 garments were developed by utilizing Zero waste pattern cutting (ZWPC) and DFD for pre- and post-consumption waste reduction. Data was gathered through questionnaires and workshops on the effectiveness of DFD execution in clothing. Findings revealed that fashion sensitive consumer was very much involved in garment redesign; moreover, DFD was a triumph in evolving a bond between the consumer and the garment. This resulted in increased the number of wears of clothing. Additionally, DFD and ZWPC together led to minimization of both pre- and post-consumer waste. The study promoted beneficial insights of using DFD for product life enhancement.

3.4 Conclusion

It is a necessity for fashion industry to adopt sustainable fashion for circular behavior to alleviate environmental impact of the fashion industry. The switch from linear model to circular model inculcates reconsideration of design and development of clothes along with their consumption and dumping. Fashion designers must focus on sustainable materials, reduction of waste, and reduction of lasting products. Guaranteeing clarity in the production process assisted wearers to make well informed preferences. Furthermore, consumers must be educated about the advantages of circular fashion so that they can reuse their garments to encourage circular behavior. Circular behavior must be adopted to reduce pre-consumer and post-consumer waste.

References

Alison Gwilt, T. R. (2011). *Shaping sustainable fashion* (1st ed.). Routledge.

Anguelov, N. (2015). *The dirty side of the garment industry: Fast fashion and its negative impact on environment and society*. CRC Press.

Bakker, C., & Schuit, C. S. (2017). *The Long View*. UNEP.

Bakker, C. A., den Hollander, M. C., van Hinter, E., & Zijlstra, Y. (2014). *Products that last: Product design for circular buisness models*. TU Delft Library.

Bhatt, D., Silverman, J., & Dickson, M. A. (2018). Consumer interest in upcycling techniques and purchasing upcycled clothing as an approach to reducing textile waste an approach to reducing textile waste. *International Journal of Fashion Design, Technology and Education, 12*(1), 1–11. https://doi.org/10.1080/17543266.2018.1534001

Bocken, N. M. P., et al. (2016). Product design and business model strategies for a circular economy. *Journal of Industrial and Production Engineering, 1015*, 1–12. https://doi.org/10.1080/21681015.2016.1172124

Bogue, R., & Bogue, R. (2007). Design for disassembly: A critical twenty-first century discipline. *Assembly Automation, 27*(4), 285–289. https://doi.org/10.1108/01445150710827069

Brass, C., & Bowden, F. (2012). Joining the dots: Connecting design to a sustainable future. In *The sustainable fashion handbook* (pp. 108–111). Thames & Hudson Ltd.

Braungart, M., Bondesen, P., Kälin, A., & Gabler, B. (2008). 'Specific public goods for economic development: With a focus on enviornment. In *Public goods for economic development. Compendium of background papers*. United Nations Industrial development Organisations.

Caro, F., & Martínez-de-Albéniz, V. (2015). Fast fashion: Business model overview and research opportunities. In *Retail supply chain management* (pp. 237–264). Springer.

Carrico, M., & Kim, V. (2014). Expanding zero-waste design practices: A discussion paper. *International Journal of Fashion Design, Technology and Education, 7*(1), 58–64. https://doi.org/10.1080/17543266.2013.837967

Chapman, J. (2005). *Emotionally durable design; objects, experiences and empathy*. Earthscan Publishing.

Chapman, J. (2015a). *Emotionally durable design objects, experiences and empathy* (2nd ed.). Routledge.

Chapman, J. (2015b). *Emotionally durable design* (2nd ed.). Routledge.

Chapman, J., & Gant, N. (2015). Prospect, seed and activate, advancing design for sutainability in fashion. In K. Fletcher & M. Tham (Eds.), *Handbook of sustainability and fashion* (pp. 74–81). Earthscan Publishing.

Claudio, L. (2007). Waste couture: Environmental impact of the clothing industry. *Environmental Health Perspectives, 115*, A449.

Derwin, E. F. (2016). *Make/Use: A system for open source, user-modifiable, zero waste fashion practice*.

Design Council. (2013, March). *Design council annual report and accounts*.

Earley, R., Vuletich, C., & Goldsworthy, K. (2010). Textiles, Environment, Design (TED): Making theory into textiles through sustainable design strategies, pedagogy and collaboration. *Textiles, Environment, Design (TED), 1*(April).

Eder-Hansen, J., Chalmer, C., Tärneberg, S., Tochtermann, T., & Seara J. (2017). *Pulse of the fashion industry*.

Fletcher, K. (2007). Clothes that connect. In *Designers visionaries and other stories* (pp. 119–132). Routledge.

Fletcher, K. (2008). *Sustainable fashion and textiles, design journeys*.

Fletcher, K. (2010). Slow fashion: An invitation for systems change. *Fashion Practice, 2*(2), 259–266. https://doi.org/10.2752/175693810X12774625387594

Fossen, V. (2017). *Business model innovation for sustainability: Towards a unified perspective for creation of sustainable business models article* (Accepted version).

Gaugele, E. K. (2008). Fashion design. In *Design dictionary* (pp. 161–165). AVA Publishing SA. https://doi.org/10.1007/978-3-7643-8140-0_108
Goldsworthy, K. (2017). The Speedcycle: A design-led framework for fast and slow circular fashion lifecycles. *The Design Journal, 20*(sup1), S1960–S1970. https://doi.org/10.1080/14606925.2017.1352714
Grose, L., & Fletcher, K. (2012). *Fashion and sustainability: Design for Change*. Laurence King.
Haines-gadd, M., Chapman, J., Lloyd, P., Mason, J., & Aliakseyeu, D. (2018). Emotional durability design bine—A tool for product longevity. *Sustainability, 10*(6), 1–19. https://doi.org/10.3390/su10061948
Hajra, M. (2021). *Design for disassembly in designing garments for circular behavior*. Master Thesis, National Textile University Faisalabad.
Hasegawa, S., Terui, N., & Allenby, G. M. (2012). Dynamic brand satiation. *Journal of Marketing Research, 49*(6), 842–853. https://doi.org/10.1509/jmr.11.0036
Hvass, K. K. (2014). Post-retail responsibility of garments—A fashion industry perspective. *Journal of Fashion Marketing and Management, 18*(4), 413–430. https://doi.org/10.1108/JFMM-01-2013-0005
James, A. M., Roberts, B. M., & Kuznia, A. (2016). Transforming the sequential process of fashion production: Where zero-waste pattern cutting takes the lead in creative design. *International Journal of Fashion Design, Technology and Education, 9*(2), 142–152. https://doi.org/10.1080/17543266.2016.1167253
Jin Gam, H., Cao, H., Bennett, J., Helmkamp, C., & Farr, C. (2011). Application of design for disassembly in men's jacket: A study on sustainable apparel design. *International Journal of Clothing Science and Technology, 23*(2–3), 83–94. https://doi.org/10.1108/09556221111107289
Johansson, E. (2010). *Slow fashion: The answer for a sustainable fashion industry?* Master's thesis, University of Borås, Sweden. Retrieved from http://bada.hb.se/bitstream/2320/6776/1/2010.9.15.pdf
Joy, A., Sherry, J. F., Venkatesh, A., Wang, J., & Chan, R. (2012). Fast fashion, sustainability, and the ethical appeal of luxury brands. *Fashion Theory—Journal of Dress Body and Culture, 16*(3), 273–295. https://doi.org/10.2752/175174112X13340749707123
Jung, S., & Jin, B. (2014). A theoretical investigation of slow fashion: Sustainable future of the apparel industry. *International Journal of Consumer Studies, 38*(5), 510–519. https://doi.org/10.1111/ijcs.12127
Jung, S., & Jin, B. (2016). From quantity to quality: Understanding slow fashion consumers for sustainability and consumer education. *International Journal of Consumer Studies, 40*(4), 410–421. https://doi.org/10.1111/ijcs.12276
Ki, C. C., Chong, S. M., & Ha-brookshire, J. E. (2020). How fashion can achieve sustainable development through a circular economy and stakeholder engagement: A systematic literature review. *Corporate Social Responsibility and Environmental Management, 27*(6), 2401–2424. https://doi.org/10.1002/csr.1970
Kwon, T. A. (2019). Why do we feel bored with our clothing and where does it end up? *International Journal of Consumer Studies, 44*(1), 1–13. https://doi.org/10.1111/ijcs.12540
Laitala, K. (2014). Consumers' clothing disposal behaviour—a synthesis of research results. *International Journal of Consumer Studies, 38*, 444–457. https://doi.org/10.1111/ijcs.12088
Laitala, K., & Klepp, I. G. (2015). Chapter 16: Clothing disposal habits and consequences for life cycle assessment. In *Handbook of life cycle assessment (LCA) of textiles and clothing* (pp. 345–365). Elsevier Ltd.
Lang, C., Armstrong, C. M. J., & Brannon, L. A. (2013). Drivers of clothing disposal in the US: An exploration of the role of personal attributes and behaviours in frequent disposal. *International Journal of Consumer Studies, 37*(6), 706–714.
Lawson, B. (2005). *How designers think* (4th ed.). Routledge.
Lewis, T. L., Park, H., Netravali, A. N., & Trejo, H. X. (2017). Closing the loop: A scalable zero-waste model for apparel reuse and recycling. *International Journal of Fashion Design, Technology and Education, 10*(3), 353–362. https://doi.org/10.1080/17543266.2016.1263364

Linton, J. D., & Jayaraman, V. (2005). A framework for identifying differences and similarities in the managerial competencies associated with different modes of product life extension. *International Journal of Production Research, 43*(9), 1807–1829. https://doi.org/10.1080/13528160512331326440

Loewenstein, G., & Ubel, P. A. (2008). Hedonic adaptation and the role of decision and experience utility in public policy. *Journal of Public Economics, 92*, 1795–1810. https://doi.org/10.1016/j.jpubeco.2007.12.011

Loschek, I. (2009). *When clothes become fashion, design and innovation systems* (1st ed.). Berg Publishers.

Maclachlan, M. (2011). *Emotional design strategies to enchance user experience and encourage product attachment*. Glasgow Caledonian University.

McDonough, W., & Braungart, M. (2002). *Cradle to cradle: Remaking the way we make things*. North Point Press.

McLennan, J. (2004). *The philosophy of sustainable design*. Ecotone Publishing.

McQuillan, H. (2019). Waste, so what? A reflection on waste and the role of designers in a circular economy. *Nordic Pulp & Paper Research Journal, 8*(8), 1–9. [Online]. Available: http://www.nordes.org/opj/index.php/n13/article/view/485/456

McQuillan, H. (2020). Digital 3D design as a tool for augmenting zero-waste fashion design practice. *International Journal of Fashion Design, Technology and Education, 13*(1), 89–100. https://doi.org/10.1080/17543266.2020.1737248

Medkova, K., & Fifield, B. (2017). *Circular design—Design for circular economy*. Conference: Smart cities in smart regions, Lahti, 2016, February.

Mishra, S., Jain, S., & Malhotra, G. (2020). The anatomy of circular economy transition in the fashion industry. *Social Responsibility Journal, 17*(4), 524–542. https://doi.org/10.1108/SRJ-06-2019-0216

Morgan, L. R., & Birtwistle, G. (2009). An investigation of young fashion consumers' disposal habits. *International Journal of Consumer Studies, 33*(2), 190–198. https://doi.org/10.1111/j.1470-6431.2009.00756.x

Morlet, A., Opsomer, R., Herrmann, S., Balmond, L., Gillet, C., & Fuchs, L. (2017). *A new textiles economy: Redesigning fashion's future*.

Moss, M. (1985). *Designing for minimal maintenance expense: The practical application of reliability and maintainability*. Marcel Dekker.

Mugge, R. (2007). *Product attachment*. Delft University of Technology.

Mugge, R., Schifferstein, H. N. J., & Schoormans, J. P. L. (2010). Product attachment and satisfaction: Understanding consumers' post-purchase behavior. *Journal of Consumer Marketing, 27*(3), 271–282. https://doi.org/10.1108/07363761011038347

Page, T. (2014). Product attachment and replacement: Implications for sustainable design. *International Journal of Sustainable Design, 2*(3), 265–282.

Peters, G., Dahlbo, H., & Perry, P. (2020). The environmental price of fast fashion. *Nature Reviews Earth & Environment, 1*(4), 189–200. https://doi.org/10.1038/s43017-020-0039-9

Ramzan, M. B., et al. (2023). Role of design for disassembly in educating consumers for circular behavior. *Sustainability, 15*(21), 15505.

Redden, J., & Galak, J. (2013). The subjective sense of feeling satiated. *Journal of Experimental Psychology. General, 142*(1), 209–217.

Rissanen, T. (1994). *From 15% to 0 : Investigating the creation of fashion without the creation of fabric waste*, pp. 1–10.

Rissanen, T. (2005). *Types of fashion design and patternmaking practice*.

Rissanen, T. (2008). Creating fashion without the creation of fabric waste. In C. Ulasewicz & J. Hethorn (Eds.), *Sustainable fashion: Why now?* (pp. 184–206). Fairchild Books.

Rissanen, T. (2013). *Zero-waste fashion design: A study at the intersection of cloth, fashion design and pattern cutting*.

Saeidi, E., & Wimberley, V. S. (2017). Precious cut: Exploring creative pattern cutting and draping for zero-waste design. *International Journal of Fashion Design, Technology and Education, 11*(2), 1–11. https://doi.org/10.1080/17543266.2017.1389997

Saha, K., & Lehew, M. L. A. (2018, June). *Durable fashion apparel: A potential sustainability solution?* In: Sustainable Consumption Research and Action Initiative (SCORAI) conference.

Sandin, G., & Peters, G. M. (2018). Environmental impact of textile reuse and recycling—A review. *Journal of Cleaner Production, 184*(March), 353–365. https://doi.org/10.1016/j.jclepro.2018.02.266

Schifferstein, H. N. J., & Zwartkruis-Pelgrim, E. P. H. (2008). Consumer-product attachment: Measurement and design implications. *International Journal of Design, 2*(3), 1–13.

Short, S. W. (2014). A literature and practice review to develop sustainable business model archetypes. *Journal of Cleaner Production, 65*(September), 42–56. https://doi.org/10.1016/j.jclepro.2013.11.039

UNEP. (n.d.). *UNEP promotes environmentally sound practices globally and in its own activities. This publication is printed on 100% eco-fibers paper, using vegetable -based inks and other eco-friendly practices.* Our distribution policy aims to reduce UNEP's car.

Walker, S. (2006). *Sustainable by design, explorations in theory and practice*. Earthscan Publishing.

Chapter 4
Present Challenges and Future Opportunities of Sustainable Retailing in the Apparel Industry

Abhijit Mukherjee and Laxmi Bharti

4.1 Introduction

A growing need to address social and environmental issues is causing a fundamental transition in the apparel business. Retailing sustainable clothing has become a significant player in this shift, and it needs to be thoroughly aware of the opportunities, problems, and upcoming trends that will configure this ever-changing industry. This chapter thoroughly examines sustainable clothing retailing and insights into the challenging environment that merchants face.

The sale of sustainable clothing presents various challenges. Retailers face challenges such as obtaining sustainable materials, maintaining supply chain transparency and traceability, controlling cost implications, and balancing sustainability goals with popular fashion trends and design paradigms. These problems necessitate creative fixes and thoughtful approaches to sustainability.

However, amid these difficulties, retailers might find bright prospects. Retailers can implement sustainable practices in their operations through creative approaches to procuring raw materials and manufacturing procedures, tactical partnerships with players in the sustainable fashion industry, sophisticated marketing and communication strategies, and the creation of environmentally friendly packaging and shipping options.

A. Mukherjee (✉)
Department of Fashion Technology, National Institute of Fashion Technology, Kolkata, West Bengal, India
e-mail: abhijit.mukherjee@nift.ac.in

L. Bharti
National Institute of Fashion Technology, Kolkata, West Bengal, India
e-mail: laxmi.bharti@nift.ac.in

S. S. Muthu (ed.), *Sustainable Apparel Retailing*, Sustainable Textiles: Production, Processing, Manufacturing & Chemistry,
https://doi.org/10.1007/978-3-031-83875-0_4

Systems regulating sustainable apparel retail must include regulatory frameworks and certification standards. Retailers that get certifications like GOTS, Fair Trade, and B Corp can show their dedication to sustainability. This industry's retail is subject to regulatory requirements and labour and environmental legislation.

This chapter will examine future technological innovations and advancements, shifting consumer preferences, market dynamics, and chances to grow the sustainable clothing business. By adopting this forward-looking mindset, retailers may use these understandings to strategically position themselves in the dynamic landscape of sustainable apparel retail.

In this context, this chapter explores a solid framework for comprehending and negotiating the obstacles, taking advantage of the opportunities, and putting oneself in a strategic position within the sustainable clothes retail industry. Retailers must adopt cutting-edge tactics and partnerships as sustainability is foregrounded in the clothing business, promoting positive change.

This chapter delves into challenges in sustainable apparel retailing, opportunities through innovative practices, strategic partnerships, and integration of sustainability with fashion design, examining the regulatory and certification frameworks governing sustainable apparel retailing and adopting eco-friendly and green packaging. This chapter also looks into the sustainable practices by small- and medium-sized retailers, emerging technologies and innovations, evolving consumer preferences, and market dynamics through meticulous analysis of a few case studies and avenues for growth and expansion within the sustainable apparel retail sector.

4.2 Sustainable Retailing and the Triple Bottom Line (TBL)

Two fundamental ideas in sustainable business practices are the triple bottom line (TBL) and sustainable retailing. In academia and corporate circles, the three universally accepted components of sustainability are represented by the TBL, which stands for People, Planet, and Profit (Su et al., 2021; Vadakkepatt et al., 2021).

Entrepreneurs who integrate economic, social, and environmental goals into their organizations and strive for sustainability in wealth-creation strategies and organizational missions are known as sustainable entrepreneurs. This idea emphasizes the dual goals of sustainable retail—preserving the environment and advancing the sustainable development of society (Todeschini et al., 2017; Su et al., 2021).

Pursuing endeavours that accomplish economic, environmental, and social goals motivates sustainable retailing. Sustainable retailers view financial objectives as a means to an end and use viable business methods to address social and environmental issues. By fusing the TBL viewpoint with the retailing process, they include sustainability into their organizational procedures, company missions, and operations (Hagan & Sporn, 2018; Vadakkepatt et al., 2021; Islam et al., 2021).

Sustainable retailers seek to address social and environmental problems while transforming the economy into sustainable ones. They see sustainable development as a business opportunity. Their strong social and ecological principles complement these retailers' entrepreneurial spirit and mentality. As change agents, they strike a triadic balance between social welfare, environmental conservation, and economic viability (Vadicherla & Saravanan, 2015; Sun, 2018).

4.3 Challenges in Sustainable Apparel Retailing

Clothes companies have made a noticeable effort to promote eco-friendly techniques in the last several years. The term 'sustainable apparel' describes clothing that combines social and environmental sustainability elements, such as organic or biodegradable materials and upholds fair trade norms without endangering the environment or the labour force. A specialized consumer market has emerged due to the global apparel industry's response to social concerns and environmental deterioration. These products and procedures offer more significant environmental and social benefits. Sustainable retailers frequently have to make trade-offs between their commitment to social, ecological, and economic goals, i.e., they have to decide between upholding social and environmental sustainability standards and promoting economic growth. Integrating sustainability into the garment industry highlights the necessity of considering the relationships between the business, the environment, people, and society. Particularly, small clothes company owners may have to choose between manufacturing and marketing eco-friendly clothing and providing goods that might not align with social or environmental responsibility, regional preferences, seasonality, or other considerations. Therefore, research on sellers' difficulties with sustainable clothing as they start and grow their businesses is desperately needed (Blissick et al., 2017; Dabija & Băbut, 2019; Urmínová & Kusá, 2021; Abbate et al., 2024).

4.3.1 Sourcing Sustainable Materials

Obtaining strategic materials in the apparel sector is fraught with opportunities and obstacles. One area where apparel makers emphasize sustainability more is using sustainable and socially conscious materials. Meeting the rising demand for garment products while addressing social issues and environmental degradation drives this trend. However, because these materials are rare and there is a chance that the supply chain could be disrupted, it can be challenging to guarantee a consistent supply (Su, 2013; Köksal et al., 2018).

Furthermore, waste streams increase due to the rising demand for garments, especially those with short product life cycles. Regulations have been implemented to limit the environmental impact and manage trash at the end of a product's life to address this.

Recycling, which incorporates materials from waste streams back into manufacturing processes, is essential to tackling these issues. Limiting the use of virgin resources lowers greenhouse gas emissions, improves resource efficiency, and lessens dependency on vital or conflicting commodities (Gam & Banning, 2011; Bhandari et al., 2022).

Even while recycling has advantages, questions persist about whether recycling particular materials is both economically and technically feasible. This is particularly true for less often recycled materials like aluminium or steel. Concerns about the quantity, quality, and timeliness of recycled materials can create trouble for manufacturers.

Additionally, governments are pushing industries to use recycled materials in their manufacturing processes. This helps the environment by increasing transparency and lowering the intricate reporting requirements for conflict materials (Goworek et al., 2012; Curwen et al., 2013).

This section examines how a manufacturer's sourcing strategy gains value from recycled materials included in the acquisition. We discuss a case in which a clothing maker chooses to use recycled products from recycling firms in addition to sourcing critical materials from virgin suppliers. We treat this as a single-period inventory problem, accounting for expenses for recycled materials, recycling amounts, and stochastic demand.

Our research addresses essential concerns such as the effect of dependencies between uncertainty, the economic advantages of employing recycled materials over conventional sourcing methods, and the social and environmental benefits of integrating recycling into sourcing strategies (Curwen et al., 2013; Su, 2013).

Lead times for acquiring raw materials present a barrier for the producer when producing garment products with a short life cycle in a newsvendor environment. To guarantee output, the clothing maker needs to order raw materials well before the season for sale. In other ways, a nearby recycling business can provide the garment manufacturer with recycled raw products with a tiny lead time. Exigent shipments of virgin products are also offered at a prime cost to satisfy any remaining demand (Rogetzer et al., 2018; Bhandari et al., 2022).

The manufacturer takes into consideration two strategies:

(a) **Standard sourcing technique (S):** This entails using emergency shipments to obtain new products at a prime cost and placing an order with a virgin product provider for new items with a lengthy lead time.
(b) **Green sourcing technique (G):** This tactic comprises, if necessary, employing emergency shipments for new products at trivial lead times, ordering new products with a long lead time from a virgin product source, and receiving recycled raw products at trivial lead times from recycling firms.

Under the green sourcing approach, emergency shipments are practiced when necessary after the recycler fulfils any missing units. With this strategy, the company may use eco-labels to educate customers about its products' sustainability while improving its public reputation and promoting environmental sustainability.

Recycled products can range in price from virgin material to less expensive than virgin material. However, some research reveals that remanufacturing a product is less expensive than creating a new one. Different research studies have produced different outcomes about the true cost of recycled raw materials with new resources.

When recycled products are used in sourcing strategies, the environment and society benefit from this sustainable approach that reduces the demand for virgin resources and diverts waste from landfills. However, the industry is also susceptible to unpredictability and dependency, such as erratic demand, recycler supply volumes, and recycled material costs. To provide a balanced and economical approach to procuring raw materials for the production of garments, these uncertainties associated with the sourcing of recycled products must be carefully considered during the decision-making process (Rogetzer et al., 2018; Bhandari et al., 2022).

4.3.1.1 Model

This section discusses a formal model for the standard sourcing strategy (S) and the green sourcing approach (G).

Green Sourcing Model (G)

The manufacturing company aims to satisfy the requirement for finished garment products, represented by the random variable D and its corresponding, presumed-to-be-invertible distribution function F(D). One unit of raw material, virgin or recycled, is required for every unit of the completed product. In addition, the recycling price C and quantity R are regarded as random variables with invertible marginal distribution functions F(R) and F(C), respectively. It is significant to highlight that D, R, and C are not presumed independent. F represents the combined distribution function of C, D, and R. The nomenclature utilized in this model is compiled in Table 4.1 (Rogetzer et al., 2018).

The chronology of events is as follows (Fig. 4.1): First before the demand for finished products is known, the manufacturer decides on the order quantity from the virgin material supplier, which delivers new raw material with a longer lead time. The purchasing price per unit of virgin material is $\mathbf{c_s}$, which is the perceived market price at the time of order placement. The actual recycling quantity and price are determined after the demand is observed. Suppose the delivered quantity from the virgin material supplier surpasses the realized demand $(\mathbf{q} > \mathbf{d})$. In that case, each unsold unit $(\mathbf{q} - \mathbf{d})$ incurs a holding cost of excess inventory, denoted as $\mathbf{c_h}$, and the sourcing process ends. If demand exceeds the ordered quantity $(\mathbf{d} > \mathbf{q})$, the company uses recycled material at a realized recycling price of $\mathbf{c}$ per unit. If the quantity of recycled material is sufficient $(\mathbf{r} \geq (\mathbf{d} - \mathbf{q}))$, the residual units $(\mathbf{d} - \mathbf{q})$ are sourced from the recycler, concluding the process. However, if the recycled material quantity is insufficient $(\mathbf{r} < (\mathbf{d} - \mathbf{q}))$, the manufacturer procures only $\mathbf{r}$ units from the recycled material supplier, and the procedure continues; any outstanding stock shortages $((\mathbf{d} - \mathbf{q}) - \mathbf{r})$ can be addressed by procuring virgin material from an emergency supplier at a premium price of $\mathbf{c_e}$ per unit (here, $\mathbf{c_e}$ is assumed to be less

Table 4.1 Notation

Abbreviation	Description
D, d	Stochastic and realized demand
F(D)	Cumulative distribution of D
R, r	Stochastic and realized recycling quantity
F(R)	Cumulative distribution of R
C, c	Stochastic and realized price per unit of recycled material
F(C)	Cumulative distribution of C
F	Joint distribution of D, R, C
q	Order quantity from the virgin material supplier (decision variable)
$\mathbf{c_s}$	Purchasing price per unit from the virgin material supplier
$\mathbf{c_e}$	Price per unit for emergency shipments
$\mathbf{c_h}$	Leftover cost per unit for excess inventory

Adopted and modified from (Rogetzer et al., 2018)

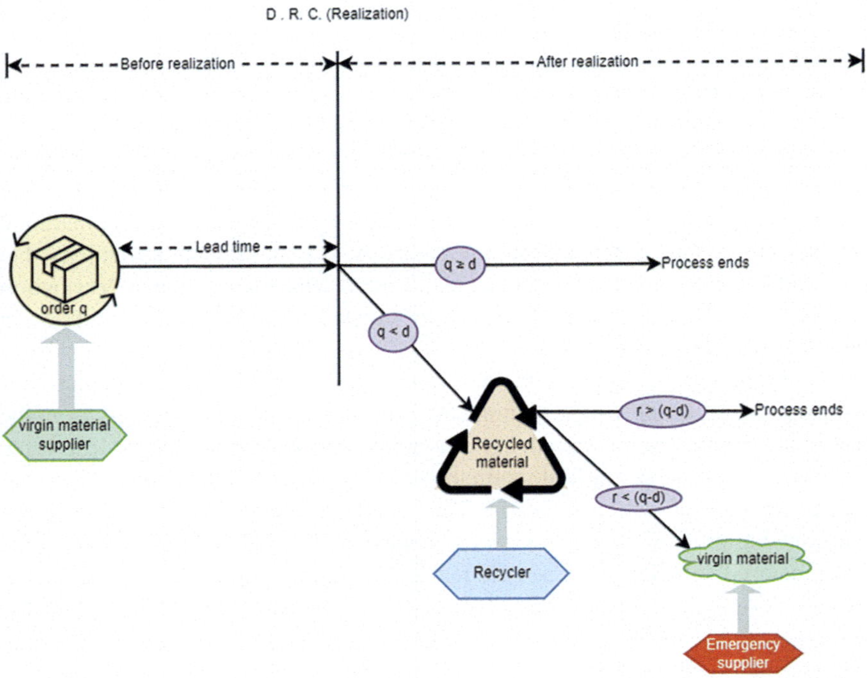

Fig. 4.1 Sequence of events in green sourcing. (Adopted and modified from Rogetzer et al., 2018)

than $\mathbf{c_s}$). Sourcing from the emergency supplier comprises higher transport costs and an additional premium, as specified in the agreement with the emergency supplier, who assures further orders at short notice and without a capacity limit.

Special Case: Standard Sourcing (S)

Materials' recycling is not a part of the standard sourcing strategy **(S)**. It entails employing emergency shipments and purchasing from the virgin material provider before the sale season. This strategy, sometimes called dual sourcing with emergency ordering or quick response with reactive supply in the literature, is used as a baseline to compare with the recycled materials-based green sourcing strategy **(G)**. To meet the arbitrary demand **D**, the manufacturer issues an order **(q)** to the virgin material supplier. If the actual demand surpasses the order quantity **(d > q)**, emergency shipments for **(d − q)** units are scheduled. Holding costs are incurred for **(q − d)** units if the realized demand is lesser than the order quantity **(d < q)** (Rogetzer et al., 2018).

Analysis

According to analysis, the following circumstances make the green sourcing strategy—which uses recycled materials—more cost-effective than the traditional sourcing strategy, which does not:

(a) **High demand variability:** Further sourcing alternatives allow for the advantageous sourcing of a bigger share of the needed amount later, which is necessary for high-demand uncertainty.
(b) **High cost of virgin material:** The manufacturer would rather source from the recycler if the price of recycled material were less than that of fresh raw material. The danger of holding costs makes recycling a more desirable alternative, even if virgin raw materials are marginally less expensive than recycled materials.
(c) **Reduction of the recycling quantity's skewness:** Reducing the recycling quantity's skewness increases the likelihood of obtaining a larger volume of recycled material.
(d) **Negative relationship between recycling price and quantity:** A company will use fewer recycled materials if the recycling price is high and there is a strong demand for recycled goods but a limited supply.

Furthermore, the green sourcing strategy has advantages for the ecology and economy. Reducing the overall quantity and supplying a portion of the whole amount from a recycler helps to reduce overall emissions.

These assessments provide businesses with decision support by helping them comprehend the impacts of uncertainty surrounding material recycling and how adopting a green sourcing strategy can result in considerable cost savings and emission reductions compared to a standard sourcing strategy.

4.3.2 Supply Chain Transparency and Traceability

Lack of visibility, trust, and efficient information sharing amongst many parties operating in silos with incompatible technology is a major problem for many supply chains. These barriers include problems with sustainability, the inability to react quickly enough to risks and disturbances in the supply chain, and distortions in demand signals like the bullwhip effect. The COVID-19 pandemic's disruption has highlighted the necessity for extended supply chain solutions, which let the whole supply ecosystem communicate, exchange information, and work together to respond to disruptions.

Businesses adopt transparency and traceability in supply chains more frequently, but many details of these strategies still need to be discovered. The distinctions between and practical application of the phrases traceability and transparency by businesses and their connection to sustainability must be better understood. Clarifying these difficulties would help establish traceability and transparency in the sector.

The literature lacks consistency and distinction in its definitions of transparency and traceability. The relationship between sustainability, transparency, and traceability is unclear, exacerbating the situation. While some academics believe that traceability is necessary for transparency, others argue that transparency lowers information asymmetry (Shen et al., 2017; Leonora et al., 2022).

Furthermore, the reasons why businesses are implementing traceability and transparency are unknown. Regulatory forces are just one of the many variables that could affect their choice. In the food business, stringent laws and regulations guarantee transparency and traceability; nevertheless, the textile and apparel industry (TAI) lacks similar controls. However, new TAI laws at the international and national levels might compel businesses to use these methods.

Environmental and social disasters are examples of sustainability challenges that push businesses to embrace traceability and transparency. Since a sizable percentage of recalls in non-food categories are related to the textile and clothing industries, product recalls, or reverse logistics may also play a role. In addition, businesses may be prompted to implement transparency and traceability by elements like supply chain management, customer demand, trademark protection, product identification, and quality assurance.

Transparency and traceability are becoming more widely used, but their effects must be thoroughly assessed. A uniform evaluation system is required to effectively inform consumers and stakeholders about the impact of these behaviours. A few measuring systems exist for transparency and traceability, such as the Fashion Transparency Index (FTI), but several are for sustainability. Nevertheless, these systems might not be thoroughly examined due to their limitations (Warasthe et al., 2020).

Blockchain technology has surfaced as a viable means of revolutionizing supply chain management, as indicated by several groundbreaking research studies. Blockchain provides an immutable distributed ledger as a safe means of logging information about accountable parties and their activities in a supply chain. This ledger records details regarding the dates, times, and locations of supply chain activities, transactions, and processes. Numerous supply chain applications have used blockchain due to its unique qualities. The four primary features of blockchain technology render it appealing for tackling obstacles in modern supply chains:

(a) **Traceability:** The sequential recording of transactions on a blockchain ensures an unchangeable historical record, guaranteeing the ledger's authenticity and making errors and manipulation efforts visible.
(b) **Security:** Blockchain's tamper-proof architecture guarantees the ledger's integrity, which only permits the addition of data and not its deletion or editing.
(c) **Transparency:** The decentralized blockchain structure allows for safe information sharing and interaction between scattered individuals.
(d) **Autonomous Smart Contracts:** Smart contracts are electronic transaction protocols that carry out a contract's provisions and enable automated execution of business transactions triggered by other transactions brought about by other transactions.

There are various kinds of blockchain networks, such as consortium networks, private or permission-based networks, and public or permissionless networks. Permissioned or consortium networks are used in the majority of supply chain management blockchain applications. The blockchain network's governance architecture and consensus mechanism determine how blocks are generated, verified, and distributed.

Utilizing blockchain technology in the supply chain has the potential to significantly influence future company models, notwithstanding its technological constraints. Blockchain's characteristics allow for secure information exchange between businesses while maintaining data ownership. In the supply chain ecosystem, blockchain can enhance trust and accountability by providing a secure and unbreakable conduit for information exchange. A blockchain can be used with current technologies like sensors, barcodes, RFID tags, sensors, and application programming interfaces (APIs) to gather and store data. This integration can increase the transparency, traceability, and sustainability of clothing goods, which can assist in addressing issues with the textile and apparel supply chain.

With its intricate and widely distributed supply networks, the textile and apparel sector stands to gain from blockchain's decentralized architecture, which encourages openness and information exchange.

Blockchain technology can facilitate materials monitoring throughout the multi-tiered garment supply network by offering a chronological transaction history. Smart contracts can promote and support moral behaviour by identifying nonconformance and enforcing sustainable practices.

Adopting blockchain technology can benefit from the prominence, influence, and standing of many players in the garment value chain. For traceability initiatives to be successful, cooperation between fibre producers, well-known brands, and retailers is essential for bridging the disparity between various supply chain sectors. Effective traceability systems depend on other vital players in the textile supply chain, including spinners, fabric manufacturers, and garment makers. Traceability systems offer a rich data environment that can enhance supply chain efficiency, inventory management, administrative procedures, and data collection, like CO_2 emissions (ElMessiry & ElMessiry, 2018; Gonçalves & Silva, 2021).

4.3.2.1 BlockChain Application

The structure and pattern of textile and apparel supply chains differ widely depending on the supply network's size, topographical spread and complexity. Apparel manufacturing includes a wide range of natural and synthetic fibres like cotton, viscose, silk, wool, polyester, etc. Despite the topographical spread of the industries and material diversity, almost entire textile and apparel supply chains mainly consist of the following value-adding stages:

(i) sources of the raw materials (farm, nature, fibre plant, etc.)
(ii) fibre manufacturers,
(iii) spinners,
(iv) fabric manufacturers,
(v) apparel manufacturers and,
(vi) brands or retailers.

Furthermore, innumerable intermediaries' merchant entities and logistics providers participate in buying, selling, and transporting textile and apparel outputs and executing specific processing activities like fabric dyeing and finishing (ElMessiry & ElMessiry, 2018).

Many industrial processes, like spinning, fabric manufacturing, cutting, sewing, and finishing, take place in less developed countries, but product design and development are primarily done in rich countries. Major merchants and brand owners frequently design and produce clothing.

Supply chain setups can use blockchain technology in several ways. Numerous blockchain projects are underway in the textile and garment industry, many using permissions or consortium networks. These initiatives aim to ensure the authenticity and sustainability of textile goods in response to the increasing demand from consumers, governments, and non-governmental organizations for sustainable practices. As the product progresses in the supply chain, blockchain technology records data on its ownership, processing of material, and movements, enabling the identification and monitoring of its origin (Fulton & Lee, 2013; Harris et al., 2016).

As a digital counterpart of the supply chain, permanent product records and related data are kept on the distributed ledger enabled by blockchain technology. In the textile and garment sector, blockchain technology has two primary applications:

(a) **Blockchain-based solutions to guarantee product legitimacy:** Many blockchain pilots concentrate on preventing fraud and ensuring the authenticity of items, especially for luxury firms that sell expensive final goods. These solutions usually follow the product from its inception to its finished form. Some high-sustainability fibre producers also use blockchain to verify their fibre and stop counterfeiting. Hugo Boss, Chargeurs Luxury Materials, Lenzing, and consortium projects like the AURA Consortium, created by LVMH, and the Arianee Consortium for digital certification of luxury goods are a few examples of these blockchain pilot projects (Ahmed & MacCarthy, 2021).

(b) **Blockchain applications to guarantee product sustainability:** Due to sustainability concerns around textile production and sourcing, numerous brand retailers and fibre manufacturers have turned to blockchain technology to guarantee garment sustainability. This is accomplished at every step of the supply chain by employing recognized suppliers and locating and confirming the provenance of the clothing. To ensure the sustainability of their products, companies and merchants like Armedangels, C&A Foundation, and Martine Jarlgaard have investigated or used blockchain technology (Ahmed & MacCarthy, 2021).

4.3.3 *Cost Implications and Pricing Strategies*

Cost considerations and pricing tactics are vital in determining profitability and competitiveness in the retail clothes sector. Retailers must comprehend these factors to make wise judgments and accomplish their goals (Fratto et al., 2006). This paper discusses some essential pricing tactics and cost consequences in the retail clothes sector.

(A) **Financial Consequences:**

(i) **Production Costs:** Labour, materials, overhead, and other associated costs are included in the price of making clothing. Variables like labour pay, material quality, and production efficiency significantly impact production costs.

(ii) **Supply Chain Costs:** Purchaser costs, production, shipping, and distribution expenses all impact total costs. Effective supply chain management is necessary to keep these expenses to a minimum.

(iii) **Inventory Costs:** An excessive amount of inventory can result in markdowns, obsolescence, and storage costs. Techniques for inventory management have two advantages: cost reduction and cost optimization.

(iv) **Marketing and Promotional Expenses:** Marketing, advertising, and promotions must be paid to attract clients. Effective marketing techniques can raise sales and brand recognition.

(v) **Infrastructure and Technology Expenses:** While investments in e-commerce platforms and point-of-sale systems increase overall expenses, they can also improve customer satisfaction and operational efficiency.

(B) **Pricing Strategies:**

(a) **Cost-Plus Pricing:** In this strategy, a profit margin is guaranteed by fixing prices by raising manufacturing costs.
(b) **Competitive Pricing:** To maintain market competitiveness, prices are set concerning those of competitors.
(c) **Worth-Based Pricing:** Instead of considering competition or production costs, prices are determined by the customer's assessment of the product's value.
(d) **Dynamic Pricing:** In this strategy, demand, seasonality, and inventory levels are considered while modifying prices to optimize revenue.
(e) **Markdown Pricing:** This tactic lowers the cost of excess or slow-moving goods to increase sales and cash flow.

(C) **Elements affecting Pricing Determinations:**

(i) **Demand in the Market:** Determining price strategies requires thoroughly understanding client preferences and purchasing behaviour.
(ii) **Pricing Strategies of Rivals:** Monitoring rivals' pricing tactics enables retailers to remain competitive and appropriately modify their rates.
(iii) **Cost Structure:** Setting competitive prices requires a thorough grasp of manufacturing and operating expenses.
(iv) **Brand positioning:** Presumably, luxury or premium brands could fetch higher costs, while value-oriented brands might concentrate on offering reasonable rates.
(v) **Seasonality and Trends:** Discounts and promotions boost sales during off-peak periods, and pricing schemes that depend on seasonal demand and fashion trends implement this (Fratto et al., 2006).

For clothing sellers, cost implications and pricing strategies are critical factors. Retailers may maximize profitability, draw in customers, and maintain a competitive advantage by maintaining strict cost control and implementing efficient pricing plans. Because of the unpredictability of demand, apparel products frequently have large markups, which causes a big discrepancy between retail price and actual cost. Clothing companies often use low-wage labour to develop notions to cut costs, which raises ethical and societal questions. Because of conventional companies' markups and high-retail pricing, consumers usually wait until sales events to make purchases. Some firms, like Oliver Cabell, Honest By, and Everlane, use cost transparency tactics and reveal their cost structure to customers to address these problems and foster trust. For instance, Everlane maintains lower markups than typical shops and discloses the whole cost of its products on its website, including expenses for labour, materials, transportation, duties, and hardware. This 'radically transparent' approach emphasizes customers' authority to know what their clothes cost. By willingly sharing cost information, brands can establish more positive consumer interactions and foster favourable views, trust, loyalty, and perceived brand values. Cost transparency can also draw attention to markup values that are lower than those of competitors, leading buyers to believe that retail prices are generally affordable. Additionally, displaying itemized costs justifies retail prices easily, which could lead to consumers believing that the prices are reasonable.

Some contend, however, that since cost transparency makes costs visible, it may discourage firms from charging exorbitant prices by preventing them from making wide profit margins. Brands that want to defend high prices must give consumers a good explanation for why a price was established the way it was since this affects how consumers view a product. To improve consumers' impressions of brand legitimacy and fairness, it is imperative to comprehend effective strategies for justifying prices through cost transparency (Nayak, 2019).

Benchmarks, such as prior brand experiences and pricing of rival brands, frequently impact consumer decision-making. According to the prospect theory, rather than focusing on an attribute's absolute value, people should assess it based on how far it deviates from a reference point. Thus, when given cost information with reference points, consumers seek gains (purchasing) and prevent losses (not buying). This section aims to experimentally examine how customers react to various framings of cost information to provide useful information for cost transparency tactics that work in the garment business (Fratto et al., 2006).

4.3.3.1 Cost Transparency Strategy

Cost transparency is the voluntary disclosure by a business of its production or service delivery costs. It may also include the exchange of cost-related data between purchasers and suppliers. Prior studies have examined the benefits and difficulties of implementing a cost transparency approach. Credibility building with clients is one of the main advantages. Furthermore, cost transparency can strengthen the bond between consumers and brands. According to a study by Kang and Hustvedt (2014), customers' opinions and trust in a company are strongly impacted by how transparent they believe it to be about its labour and production practices, which affects their intentions to purchase. A cost transparency strategy enables consumers to create emotional assessments of brands and products and to evaluate pricing justice by providing cost information. Furthermore, this strategy has been successfully executed as a cutting-edge marketing technique to highlight firms' moral and responsible business practices (Jung et al., 2020).

The difficulty in justifying price premiums, however, is a major obstacle to providing transparent cost information, as doing so may make it more difficult to set high retail prices with big margins. Due to the fast degradation of product value and the erratic and unpredictable nature of demand, higher markups are often needed, especially for clothes companies. As a result, many merchants provide discounts both during and after the season. As such, clothing companies might be reluctant to disclose their comparatively large markups.

4.3.3.2 Retailers and Prices

Retail Channels

Networked businesses that work together to make goods or services accessible for usage or consumption are referred to as retail channels. The retail garment sector generally employs two general competitive strategies: distinctiveness and cost leadership. Being the lowest-cost manufacturer is the main goal of cost leadership, which calls on economies of scale, access to raw materials, and proprietary technologies. Conversely, distinctiveness aims to be unique in the market through marketing, product, or channel. Neiman Marcus for distinction and Wal-Mart for cost leadership are two examples of businesses using these tactics (McCormick et al., 2014).

Retailer Power

Retailer power is the capability of a channel participant to influence another to enhance its behaviour. Large retailers frequently wield great influence in the apparel sector since products are supplied directly from manufacturer to retailer to customer. Several variables, such as influence over national brand price promotions, information technology developments, changes in retail formats, and desire for shorter lead times, have contributed to this power shift from producers to retailers.

Retailer power is demonstrated by control over product assortment, price, inventory, and store brand development. Clout retailers in the market, such as Wal-Mart, have compelled suppliers to deliver goods on a consignment basis (Fratto et al., 2006).

Pricing Strategy

Retailers typically use established policies as the foundation for their price decisions. A set percentage markup is sometimes applied to costs by those who employ a rules-based approach; for example, a keystone markup yields a 50% markup on the retail price. Fashion merchants often apply a pre-set percentage reduction to items in the store for a certain number of weeks, followed by an additional markdown later. These criteria also apply to markdowns. One more facet of the rules-based method is setting expenses for goods that are higher, lower, or equal to those of rivals.

Market characteristics, chain size and positioning, shop size and category assortment, brand equity, and consumer sensitivity to price adjustments are some of the elements that affect pricing tactics in the retail sector. Aligning pricing strategies with assortment overlap and longevity is essential for effective pricing strategies. This is especially true for perishable or time-sensitive product categories, such as clothes (Dabija & Băbut, 2019).

The rise of retail giants like Walmart has altered the industry's pricing methods, as competitors now have to differentiate themselves by offering higher-quality items or services or engaging in direct price competition. Manoeuvring through the ever-changing terrain of the garment retail market requires a thorough awareness of retail channels and the elements that impact pricing tactics (Fratto et al., 2006).

Recent trade news articles show merchants have slowly implemented more complex pricing models. They frequently price the products only according to cost and mostly rely on 'Excel' for assessments. Because the rules-based method is easy to compute and implement—especially in a chain of multiple stores—retailers find it enticing. Furthermore, pricing in comparison to competitors aids in maintaining a retailer's price image.

The primary flaw of these rule-based methods is that they do not always represent the best deal or markdown. Rather, they follow previous trends, such as the prior year for clothing or the last few weeks for basic products. This information is frequently skewed by marketing and out of date. Retailers might not be able to assess the effect of a Valentine's Day campaign on sales, for instance, without price optimization tools and knowledge about prior promotions.

The system-wide nature of the decisions made by merchants about merchandise pricing and markdowns is another problem. For instance, even though the demand and selling seasons for sweaters may fluctuate significantly, a local department store chain may offer the same markdown on sweaters in Texas and New England. In Miami, where a large demand is there for black beans owing to the Latino population, a grocery chain might price black beans similarly to Tallahassee, where they are more of a novelty.

Most retailers are reluctant to implement varied prices in stores in the same vicinity because they worry about confusing customers or raising doubts about their honesty. Differential pricing, however, may result in increased gross margins and more accurate inventory control at various locations, especially remote locations.

Thirdly, rule-based systems have the drawback of customers waiting for sales to make purchases, as they possibly infer from past experiences when products will be put on sale. This practice may impact retailers' profit margins.

Despite these issues, retailers still need to rely on rule-based strategies due to the scarcity of data, the demand for more complex analysis, and time restrictions. Due to the vast quantity of products they carry and the timing of studies, most merchants find this strategy impracticable, even if some can test new things to find the best prices. However, when launching a new menu item, companies like Cheesecake Factory may discover that experimenting with different prices in various markets helps them determine the most profitable price (Fratto et al., 2006).

4.3.4 Balancing Sustainability with Fashion Trends and Design

Although this section aimed to learn how people saw the integration of sustainability into fashion design, another important goal was to quantify and identify the differences in participant perspectives and gather participant experiences through open-ended comment sections. A five-point scale varying from ‘strongly disagree’ to ‘strongly agree’ indicates the obstacles respondents found when attempting to adopt sustainable fashion. The degree of difficulty for each category is indicated.

Most participants thought that the biggest problem was incorporating sustainability into design practice. Other significant challenges included a lack of knowledge on incorporating sustainable design and design process guidelines. Restricted options for materials and design procedures were also considered to be major obstacles.

Some respondents disputed the claims that ‘I don’t know this area’ and ‘My decision does not influence the design processes’. Although most respondents appeared to be aware of sustainable fashion design, it was difficult for them to put the right solutions into practice due to sustainability’s complexity. ‘Limited material choice and procedure’ and ‘not my company or my design strategy’ have substantial standard deviations, indicating respondents’ different opinions about these claims.

This section incorporates open-ended replies to gather diverse perspectives on these difficult subjects. Many participants acknowledge that incorporating sustainability into fashion design faces both internal and external challenges. One fashion sector respondent, for instance, raises the issue of consumers’ ignorance of sustainable design or products.

In addition, respondents mentioned internal obstacles. Some respondents mentioned internal challenges to sustainability in fashion design, even though most respondents thought their decisions were significant and impacted the design process. As genuine design processes are primarily driven by fashion trends related to aesthetics, sustainability is not a top concern in fashion design, according to one experienced fashion designer.

These results imply that the fashion industry and consumers must address sustainability challenges in fashion design. Growing consumer knowledge may increase demand for eco-friendly goods, pushing the fashion industry to use eco-friendly design techniques. However, if sustainability is not a top priority for their company’s culture, designers could find it difficult to include it (Nayak, 2019).

4.3.4.1 A Holistic Designing Approach for Sustainable Business Development in Fashion Value Chains

Developing a lasting competitive edge in businesses requires the strategic adoption of independent or concurrent designing of products, processes, and supply chains. By integrating early supplier involvement (ESI), concurrent engineering (CE), quality management, and customer involvement, these practices adhere to the principles of Three-Dimensional Concurrent Engineering (3DCE). They also match supply chain design attributes to product and process design requirements. Financial performance is improved as a result of this integration. By tackling underlying issues linked to non-sustainable performance across all strategy vectors, 3DCE offers a comprehensive perspective. While assessing economic viability and sustainability has been the primary emphasis of organizational, functional diagnostics along the 3DCE to identify distinctive competencies, simultaneous design of the product, process, and supply chain has been suggested as a way to improve conventional new product development outcomes. These results include a shorter time to market, lower costs, and higher client approval. Along with the three sustainability pillars, competitive priorities like enhanced product innovation, decreased lead time and time-to-market (TTM), delivery performance (speed), cost minimization, and reduced relationship risk are measurement characteristics for determining organizational performance (Pal, 2014).

Design for Sustainability in Design and Product Development

To lessen the burden on natural capital (water, chemicals, energy, raw materials), companies must adopt sustainable product design initiatives like ecologically intelligent design, product remanufacture and reuse strategies, recycling and material transformation strategies, and green product design. Design for Sustainability (DfS) is categorized into two distinct clusters: 'eco-centric design', which is strong in sustainability, and 'techno-centric design', which is weak in sustainability. The fashion industry must adhere to the principles of economic, social, and ecological sustainability, eliminate negative environmental impact through skilful design, reduce the use and effects of non-renewable resources, use more recycled and reusable materials, and foster a sense of connection between people and the natural environment to realize the strategic role of sustainable product design fully (Moon et al., 2013; Fulton & Lee, 2013).

The fashion industry uses life cycle assessment (LCA) as a tool to measure the consequences of all the processes involved in creating and utilizing garments. Manufacturers and brands utilize life cycle assessment (LCA) to convey their business and product impacts, pinpoint areas of environmental impact, and optimize product design and operations. With its integration of cradle-to-cradle thinking, the cradle-to-cradle apparel design (C2CAD) method has grown in popularity. This strategy aims to create models for sustainable clothing design and manufacture. Manufacturers of textile products, including Nike, Designtex, and Shaw Industries, have effectively created more ecologically friendly fabrics and products by utilizing the C2CAD paradigm.

In fashion value chains, demand ambiguity frequently results in overproduction, which impairs supply chain efficiency and long-term economic viability. Reducing environmental impact requires using sustainable product design techniques like creative pattern creation and zero-waste clothing design. Creating multipurpose clothing, utilizing zero-waste patterns, and utilizing quick prototyping methods are some tactics for cutting fabric waste and increasing efficiency (Moon et al., 2013; Pal, 2014).

Design for Sustainability in a Closed Loop

Remanufacturing, recycling, reverse logistics, and redesigning are crucial components of sustainable business development (SBD) to close the supply chain operations loop. Fashion gear can continue to be used in various ways after being used by consumers. Clothes and textiles can be brought to recycling banks run by groups such as Oxfam, Traid, Salvation Army, etc. In these banks, roughly half of the textiles collected are recycled, and the other half are repurposed. When producing with recycled textile material instead of new materials, carbon emissions can be greatly reduced. For instance, consuming 1 kg of recycled clothing can cut 3.6 kg of carbon emissions, 6000 L of water, 0.3 kg of fertilizer, and 0.2 kg of pesticides (Fulton & Lee, 2012).

The Stena Metall Group's study on recycling clothing with metal components, like bras with metal brackets or jeans with rivets and zippers, is one creative example. Textiles undergo pyrolysis, which converts them into a gas that cools and creates oil that may be recovered, leaving the metal components intact. However, this procedure can be costly, and handling different kinds of materials and using a lot of fibre blends can lead to major bottlenecks. Despite these obstacles, several businesses—like H&M's Garment Collecting and Klättermusen's deposit system—have voluntarily allowed consumers to return clothing for recycling and reuse (Yang et al., 2017).

Initiatives like the 1990 Recycling of Carpet Materials (RECAM) project, while technologically effective, failed to be economically viable and profitable. Recycling requires technological advancements to recover longer fibres from worn textiles. However, recycling is financially feasible and lowers the energy consumed in production. Another strategy to close the loop is to reuse the approximately $1 billion worth of clothing produced annually into second-hand clothing, which supports social sustainability and jobs in less developed countries. From an eco-friendly standpoint, reverse logistics optimizes the movement of goods for recycling, remanufacturing, or disposal, resulting in advantages for the environment and the economy (Pal, 2014). Upcycling has become more popular to give unwanted goods new life and has opened up commercial options for smaller fashion labels in particular. Creating a closed-loop value chain and guiding the fashion sector towards sustainability entails upcycling used apparel, sophisticated trash recycling, and new product and process design viewpoints.

4.4 Opportunities for Retailers

India's recent dramatic social and economic transformations have given foreign retailers access to a substantial consumer base. The average GDP growth rate of 4.5% between 1997 and 2007 indicates that the Indian economy is in good shape. By 2050, India's economy is expected to rank third in the world, behind China's and the USA's. India is ranked in the top five countries for retail apparel based on market size, growth potential, customer wealth, and readiness.

Trade liberalization is credited with this prosperous time. Foreign direct investment (FDI) has been promoted by the Indian Government in every economic and profitable sector since 1991. In the retail sector, relaxed foreign direct investment (FDI) restrictions were implemented in 2006, allowing 100% ownership in wholesale cash and carry products and 51% ownership in single-brand retailing. Accordingly, the Indian retail sector is prognosticated to expand at a 13% annual rate, from $322 billion in 2006–2007 to $590 billion in 2011–2012. Consequently, the textile retail sectors are predicted to expand 7.8% annually, ending at $39.4 billion by 2013, a 76.7% growth from 2006 (Mann & Byun, 2011; McCormick et al., 2014).

4.4.1 Communication and Marketing Strategies to Promote Sustainability

4.4.1.1 Green Communication

All corporate and marketing communications about sustainability are included in the category of sustainability communication, including CSR communications that include social responsibility, environmental preservation, and financial achievement. There are numerous terms for sustainable communication, such as green communication, global responsibility communication, social responsibility communication, and corporate sustainability communication.

In the modern age of digital communication, the web has emerged as an indispensable medium for reaching out to stakeholders, and corporate websites are a must for presenting sustainability goals and company statements in an eye-catching and straightforward manner.

Companies now use sustainability reporting to inform stakeholders about their performance and commitments. Large European corporations have been obliged to provide this information in their Non-Financial Reports since 2014. The European Commission has further enhanced the criteria for sustainability reporting in 2021 with recent revisions.

Traditional media is being replaced by digital platforms, which impacts CSR, sustainability communication, and corporate communication. Building relationships with stakeholders and improving a company's reputation depends heavily on online communication.

Advances in digital communication, such as social networks and interactive media, have increased conversations on social and environmental issues and brought attention to sustainability and corporate social responsibility (CSR). However, these advancements have certain risks, as misinformation or a lack of transparency can result in communication breakdowns.

Customers are now significant stakeholders, which increases the demand for sustainable and corporate social responsibility messaging. The Internet and social media have significantly facilitated the sustainability movement, influencing consumer attitudes and purchase patterns.

Fashion firms still struggle to communicate about their CSR and sustainability initiatives properly, and many are still unable to share their practices. A company's inability to regulate suppliers and fear of criticism are two common obstacles to external sustainability communication (Dal Mas et al., 2022).

Transparency and consistency in messages are essential to effective sustainability communication since they affect consumer attitudes and understanding. Involving consumers in sustainability communication may result in more sustainable consumption patterns.

In general, sustainability communication aims to inform customers and influence reforms within the fashion sector. A developing trend is using corporate websites to convey sustainability objectives, although most enterprises have active websites for e-commerce and brand distinction (Sanmiguel et al., 2021).

4.4.1.2 Green Marketing

A new series of obstacles for organizations to overcome for survival has been brought about by stakeholders' increased knowledge of the negative consequences on the environment. Since sustainability is a critical environmental strategy in corporate decision-making processes, more and more businesses are adopting a holistic approach to it. Driving this change is the need to preserve competitive positioning, satisfy external legitimacy requirements while protecting the environment and meet the increasing demand from consumers for eco-friendly products.

This strategy, sometimes called 'green marketing', entails ingeniously incorporating sustainability concepts into marketing procedures. This concept has been referred to by several names, including ecological marketing, environmental marketing, and sustainable marketing; it is generally known as 'green marketing'.

Green marketing has seen various stages of development, encompassing ecological, environmental, and sustainable phases. It has its roots in social marketing and is mostly found in management literature because of its special capacity to combine ecological concerns with corporate operations. Green marketing is a corporate strategy that includes a comprehensive marketing process to produce goods and services while considering ecological concerns and improving financial performance.

The primary objective of green marketing, which seeks to align an organization's marketing tactics with the current environmental situation, is adopting greener practices. In the twenty-first century, implementing green marketing is not only a decision—it is also an essential step towards addressing environmental problems and enhancing an organization's reputation for accountability.

Various factors drive green marketing development, including economic opportunities, social obligations, legal constraints, competitive pressures, and cost and profit considerations. Companies use green marketing to enhance their brand, manage environmental concerns, seize business opportunities, forge long-term competitive advantages, and boost customer satisfaction.

In addition, corporations are now placing equal importance on sustainability performance and financial returns. To satisfy important stakeholders, sustainability performance entails concurrently enhancing environmental, social, and economic performance. It is regarded as a modern corporate strategy to manage and resolve environmental challenges to boost financial performance.

Scholars and practitioners are beginning to embrace the idea of sustainability, which highlights the value of social progress, environmental preservation, and financial success. Preserving harmony between short-term benefits and long-term consequences is crucial since sustainability is defined as an evolution that fulfils present demands without endangering the ability of upcoming generations to meet their requirements (Ara et al., 2019).

4.4.2 Developing Eco-Friendly Packaging Solutions

Green manufacturing, packaging, and shipping are just a few techniques enterprises have been compelled to implement by government requirements. These actions have a substantial impact on both environmental and economic performance. Through sharing best practices and reducing environmental hazards during the production process, suppliers are essential in helping manufacturers maintain these green practices. Fostering ecological awareness requires organizations to transform their supply chains into green ones. Green Supply Chain Management (GSCM) techniques can enhance ecological equilibrium and foster a salubrious atmosphere by mitigating the unfavourable ecological impacts of business operations. The Government can support and encourage green practices in several ways, such as by purchasing, facilitating, or regulating them (Yang et al., 2017; James & Kurian, 2021).

4.4.2.1 Sustainable Packaging

Packaging has long been used to protect products and prevent losses from happening from the point of manufacture to the final consumer. It is crucial to marketing because it assists brand owners in creating a distinctive niche for themselves in the marketplace. The packaging design must draw in customers, deliver a message, pique their interest in the goods, and finally make a sale. The goal of packaging design should be to optimize productivity and logistics. 'Packaging logistics', according to Saghir (2002), is the process of organizing, putting into practice, and managing a coordinated packaging system aiming to maximize consumer value, sales, and profit through the safe, assured, efficient and efficient handling, transportation, distribution, storage, retail, consumption and restoration, reuse, or disposal of goods. Companies can save 3%–5% on expenses for the supply chain by implementing green packaging strategies. There will be a shift in consumer behaviour towards eco-friendly products as environmental worries increase. In developed countries, environmental concern is a reliable indicator of green purchasing. Using sustainable practices improves a company's reputation and brand image while conserving resources and reducing its carbon footprint (Yang et al., 2017; Naidoo & Gasparatos, 2018).

Green Packaging

Using environmentally friendly and biodegradable materials to package products is the core idea behind green packaging. With the increasing popularity of green supply chain management, enterprises worldwide are being compelled to adopt more ecologically friendly packaging methods. Some methods include minimizing packing to achieve an appropriate weight-to-volume ratio and using recyclable and biodegradable materials. The primary criteria used to assess current packaging practices are the expansion of e-commerce, the need for easy packaging and improved consumer experiences, and the growing push to reduce waste and pollution. The bulk of goods delivered have expanded owing to the growth of e-commerce, necessitating longer and more intricate logistical networks for parcels. Now, at all times, durable, environmentally responsible packaging that boosts brand awareness and offers a satisfying shopping experience is essential. When consumers' dissatisfaction with unsuitable and useless packaging escalates, the need for alternatives is in high demand. The want for more sustainable and eco-friendly packaging results from the growing awareness of consumers on the adverse impacts of plastic and other non-biodegradable materials. However, if consumers have to pay a little extra for sustainable packaging, it matters a lot for a cleaner environment than any massive move (Naidoo & Gasparatos, 2018).

Companies need to create long-term sustainable packaging solutions to address this shift in customer perception. Data science and artificial intelligence developments can facilitate companies' creation of precise master data for product dimensions and in optimizing packaging. Collaborative robots are also employed to streamline the packing process. Businesses worldwide are looking into alternative packaging materials like fibre-reinforced stretch films, pallet locks, lids, and trap systems to utilize less plastic shrink wrap.

In the supply chain, closed-loop logistics—which uses circular and reusable packaging—is becoming increasingly popular as a substitute for short-life plastic. Using circular and reusable packaging solutions was noted as a top concern in

DHL's customer survey. Businesses are looking into cutting-edge business strategies to get consumers to return packaging for reuse, like RePack's closed-loop packaging technology. Additionally, subscription models for reusable packaging are being investigated (James & Kurian, 2021). In general, customer perceptions of environmentally friendly packaging impact their propensity to buy, underscoring the significance of sustainable packaging in the current marketplace.

4.5 Regulatory and Certification Frameworks

The definition of corporate responsibility (CR) needs to be clarified and open up for discussion. According to free market proponents, CR aims to maximize profits while following the law. However, a more thorough understanding of CR includes environmental stewardship, labour rights advancement and respect (as described by the International Labour Organization, ILO), and human rights protection (as stated in the United Nations Declaration of Human Rights, UDHR). This variety of viewpoints draws attention to how CR might be interpreted.

This section examines how players interpret and react to critical race theory (CR), highlighting the intricate relationships between CR actors in different institutional and normative settings. The desire for ethical business practices prioritizing worker rights and ethical treatment drives transnational corporations (TNCs) to address labour-related concerns in their supply chains. However, weak governance, corruption, and rule of law frequently obstruct efficient national laws and governance systems in nations where TNC suppliers are based, leaving a regulatory vacuum. Some market participants have taken advantage of this regulation void to increase earnings. As a result of civic society's realization that the Government cannot adequately handle these problems, TNCs have been tasked with guaranteeing worker protection (Lamrad, 2013).

4.5.1 Sustainability Certifications

By providing an external indication, sustainability labelling allows brands to highlight unobservable features or practices of their products, filling in knowledge gaps and influencing consumer behaviour. This labelling serves as a 'nudge' or component of the choice architecture intended to encourage consistent customer behaviour. Nudge tactics are not new; they have been applied in food and automobile manufacturing, for example, but their usefulness in connection with sustainable fashion is up for discussion. Using identifiable certifications to boost customer trust and environmental concerns is one strategy to escalate the effectiveness of sustainability labelling 'nudges' because trust has a major impact on purchase intention. Customers want openness in the sources of products, how they are made, and how they affect the environment, yet they frequently need more information to evaluate these aspects without explicit brand communication.

When a reliable third party certifies a trading partner's skills, it can be a helpful tool for communicating sustainability. Reputable sustainability certification has advantages for businesses, clients, and suppliers. On the other hand, deceptive labelling, or 'greenwashing', misleads customers and erodes confidence.

By utilizing sustainability as a marketing tactic, fast fashion shops have muddied sustainability certificates, which were meant to promote openness and counteract misleading claims in the fashion sector. Consumers see sustainability certifications as indicators of ethical product practices; nevertheless, the fashion industry's use of several certifications, labels, and assessment systems has confused consumers and weakened brand messaging. One reason for the ongoing discrepancy between consumer opinions and behaviours when it comes to buying garments is the absence of useful information regarding fashion consumption (Baker & McNeill, 2023; Oelze et al., 2020).

4.5.1.1 Certification Forms

Fashion is complex consumer goods frequently chosen for purposes other than utility, with each article of clothing expressing several informational signals. Though essential, limited research has been done on sustainability cues in fashion because of their intricacy. Even though more and more customers understand the need to make sustainable decisions, many still find it challenging to adjust their behaviour. This points to a crucial era of transition for decision-making and sustainability.

Fashion sustainability labelling can be generically classified into organic, animal, social, environmental, holistic, and recycling categories. Global Organic Textile Standards, SA8000 Social Accountability International, Worldwide Responsible Accredited Production, Fairtrade, and B Corp by B Lab are the most popular third-party certifications. However, the rise in popularity of brand-created certificates, like H&M's 'Conscious Collection', is confusing consumers.

Labels can be ordinal/ratio (communicating a specific score), nominal (like brand graphics), or binary (like the Fairtrade symbol). The current deluge of labels and lack of comparability make it difficult for consumers to find more sustainable explanations. Similar to the 'traffic light' system on food items, a standardized, single-label strategy would streamline operations and promote sustainability in the clothing industry.

Third-party accreditation is essential to providing customers with accurate information; however, the absence of formal certification in the fashion business results in confusion. The Sustainable Apparel Coalition's Higg Index, which offers a common instrument to assess social and environmental effects, has tackled this problem. However, there are concerns with how the elements are weighted (Baker & McNeill, 2023).

4.6 Case Studies and Best Practices

4.6.1 Case Study 1

This case study looks at people tree, a small- and medium-sized enterprise that sources organic and fair trade cotton apparel from developing nations. The company, which operates in Italy, Great Britain, and Japan, is notable for its openness and gives clients access to information about its suppliers. This openness demonstrates the business's moral righteousness, which is essential for drawing in customers.

People Tree, founded by Safia Minney, stands between the corporate and nonprofit sectors by prioritizing traditional skills and producers over financial gain. Despite this emphasis, the business has expanded its market share and sales turnover in the fair trade apparel sector. People Tree's commitment to the environment is evident in its product development; it values eco-friendly processes, fair pricing, rural development, traditional skills, and transparency.

The company sources from over 70 Fair Trade groups in 20 developing countries to benefit underprivileged producers and communities. People Tree's approach to social responsibility includes financing educational and vocational training institutes and demonstrating a comprehensive commitment to sustainability.

People Tree wants to influence people's perceptions of ethical apparel by cooperating with popular fashion companies, employing modern marketing tactics, and interacting with celebrities. The brand believes in ethical fashion in the coming years ahead and underlines the significance of long-term commitment and investment in Fair Trade standards.

Overall, People Tree's business strategy combines commercial and social objectives to achieve a triple bottom line considering financial, social, and ecological benefits. This is compatible with the general trend of big shops implementing CSR policies, indicating a change towards more sustainable practices in the fashion industry (Goworek, 2011).

4.6.2 Case Study 2

Nike's Sustainable Business: A Sustainable Planning Strategy External Pressures on Nike

(a) **Environmental Pressure:** The environment is a primary pillar of the Triple Bottom Line. Human needs are limitless, while natural resources are finite. The fast rise of the human population creates a significant need for natural resources. If this trend continues, the shortage of natural resources will have an adverse impact on future generations' welfare. So, a major issue is how to preserve the planet's resources. The world economy has expanded quickly during the last two decades, facilitating the growth of the fashion sector. The demand for agricultural productivity grows with the increasing rise of the fashion sector. According to the Committee (2018), the anticipated global cotton consumption in 2017/2018 increased by 11%, or approximately 25.4 million tons, over 2016/2017. Meanwhile, the International Wool Textile Organization (2015) reports that 1.160 million kg of clean raw wool was generated from 1.163 billion sheep in 2015 to meet global demand. In 2015, the agricultural industry was put under strain, and the ecosystem suffered due to the high consumption of natural resources (plant-based and animal fibres). Overgrowth and grazing cause soil erosion and the loss of organic carbon—both are detrimental to the agriculture industry. In 2015, the agricultural business was stressed, and the ecosystem suffered due to the high consumption of natural resources (plant-based and animal fibres). Overgrowth and grazing resulted in soil erosion and organic carbon loss, damaging agriculture. Water contamination is becoming increasingly significant as the population and activity levels rise. Approximately, 80% of human-generated wastewater is dumped into waterways without being treated. Meanwhile, the apparel business is highly water-intensive and polluting (Fung et al., 2019). As previously said, cotton usage for textiles and clothing is large, resulting in a strong demand for raw cotton. Pesticides and preservatives are widely utilized to secure the stability of the cotton supply, accounting for 24% and 11% of global usage, respectively. At the same time, these harmful chemicals are washed into waterways, causing considerable harm to ecosystems. Nonetheless, the dyeing and finishing processes in fashion textile production account for 20% of global industrial water pollution. To address these issues, fashion companies in the fashion apparel supply chain could deplete their environmental footprint, which benefits the ecosystem and prevents the planet's environment from deteriorating. It is possible to save natural resources by properly structuring the retailer's business model and supply chain. This can help both the environment and the sustainable fashion apparel industry. Competing bands, like Adidas, have attempted to contribute to decreasing water pollution in the production process. They invested in a new dyeing process known as 'Air-Dye'. Compared to typical dyeing processes, utilizing Air Dye requires only 5% of water and 14% of energy for production, while greenhouse gas emissions are 84% lower (Hepburn, 2015).

(b) **Social Pressure:** The Triple Bottom Line framework recommends that businesses implement social responsibility measures because people's well-being is a key sustainability focus. Fashion companies should prioritize labour equality and participate in fair business practices that benefit the community and society. For example, fashion retailers should conduct fair business with their stakeholders. By giving their upstream stakeholders enough money, stakeholders may pay fair wages and offer adequate working conditions for their employees. Citing worldwide fashion business data, a considerable expansion in employment in the fashion industry has been recorded during the last 30 years. The workforce engagement in the garment manufacturing, textile and textile-clothing industries registered an increase of 71%, 68%, and 69%, respectively, during the period from 1990 to 2014—amounting to about 3.5% of the worldwide workforce. The above employment figures demonstrate the gravity of the fashion industry's ethical performance, which can affect worldwide poverty levels. However, it has been stated that several large fashion corporations are facing ethical concerns in manufacturing and production (Caniato et al., 2012). Adidas, Benetton, and Levi Strauss are accused of ethical difficulties during production and manufacturing. Moreover, the elimination of gender discrimination, along with the restoration of equality for women, has gained a severe and long-standing concern for decades. Even though empowering women and girls has benefitted the economy in several ways, women continue to confront uneven job possibilities in certain areas. It has been claimed that males earn 27% more than women for the same job. To address this issue, the fashion sector provides work options for women in Turkey. The International Cooperation and Development Agency (SIDA), the United Nations Development Programme (UNDP), and the Regional Development Administration of Southeastern Anatolia Project (GAP) were formed later. More than 4300 women in Southeast Anatolia may earn money by participating in manufacturing and other project-related activities thanks to this collaboration (Helmore, 2018).

(c) **Economic Pressure:** The economic pillar of the Triple Bottom Line paradigm focuses on how firms' business progress affects the global economy. It advises that businesses implement long-term profit methods that benefit both their firm and the economy. With this win–win situation, future generations may enjoy a strong economy with steady growth. The average number of clothes bought per capita in 2018 climbed by 60% over 2000. However, only 60% of the purchased clothes were utilized, with 40% never worn before disposal. According to the population growth rate, garment consumption is predicted to rise from 62 million to 102 million tons by 2039, implying that more clothes will be purchased and disposed of. Furthermore, the approaching rapid fashion trend will boost fashion consumption even further. The expansion of the fashion garment industry and its associated demand has led to an increase in pollution. Fashion firms should evaluate their business strategy and marketing techniques while making business choices and focusing on company growth. Fashion merchants should promote slow fashion to help the fashion industry become more sustainable. Slow fashion has three basic components:

(1) **identifying the value of local resources,**
(2) **ensuring manufacturing transparency, and**
(3) **prolonging the product's useable life.**

Everlane, a U.S. apparel brand, completely incorporates the notion of slow fashion into its business strategy. It strives to provide great product transparency to its customers. We can see the clear cost breakdown of each product on its website, which includes material price, hardware price (sundries), labour cost, tariffs, and transportation cost (freight) (Fung et al., 2019). Furthermore, the website displays the source of manufacture for each relevant product. Everlane refers to this business strategy as 'Radical Transparency'. Everlane can develop a transparent and sustainable supply chain by carefully selecting factories. Also, material selection may extend the life of garment goods. Furthermore, Everlane aims to educate its customers on ethical consumption. On the contrary, such a business model has the prospects to improve fashion sustainability while also promoting a sustainable corporate sector. The factories with a high ethical score of compliance can be highlighted on the Everlane website, perhaps attracting other companies to contact them for manufacturing (Jung & Jin, 2014).

4.6.2.1 Internal Strengths and Weaknesses and Corresponding Action Plans of Nike

(i) **Economic:** Nike's revenue increased 6% to $32.4 billion in 2016, intending to grow further in 2017. To do this, the organization needs significant resources for long-term development. The management anticipates that significant energy resources would be required, but shifting energy costs due to the oil crisis may affect growth. Nike has over 75 distribution locations worldwide, leading to increased energy use and carbon impact. To overcome this, Nike pursues low-cost renewable energy sources. Its logistics campus in Belgium uses 100% renewable energy, with solar panels providing 35% of the Taicang logistics centre's electricity, lowering expenses and carbon footprint.
(ii) **Environment:** Pollutants and waste are generated throughout the manufacturing process of clothes, including Nike. Nike aims to cut its environmental impact in half in 2016, setting waste reduction objectives for its factories and suppliers. To reduce water use, Nike created the 'Nike Water Minimum Programme,' which gathers wastewater treatment data to improve water efficiency and build water recycling infrastructure. Nike also intends to eliminate hazardous chemical discharge by employing the AFIRM Group's restricted substance list (RSL) in 2017 to test and analyse chemicals used in garment manufacturing, ensuring compliance and reducing hazardous chemical consumption.
(iii) **Social:** Nike attracted criticism for running sweatshops. To address this, Nike works with the Fair Labour Association (FLA), Better Work, and other third-party groups to create the Sustainable Manufacturing and Sourcing Index (SMSI) (Fung et al., 2019).

Nike's products cannot be manufactured in factories rated below the bronze level. Nike also faced a gender discrimination lawsuit, which prompted modifications in its human resources policy, including the 2016 establishment of the 'U.S. Family Care Benefit Programme' and new HR training and monitoring systems for 2019 to encourage equality and fair treatment.

4.6.2.2 Nike's Product Design Strategy and Stakeholder Implementation

(i) **Economic:** Nike developed product designs to boost company growth, such as the 'flyknit' technology for footwear, which reduced production complexity and human resource requirements. This enabled a faster response to market needs and trends.
(ii) **Environmental:** Nike strives for zero waste in product design, employing the Sustainable Materials Index (SMI) to guide material selection and collaborating with suppliers to promote sustainable materials like organic, recycled, or BCI cotton.
(iii) **Social:** Nike incorporates sustainable living principles into product design to encourage healthy and active lifestyles, providing designers with sustainable information from design concepts to material choices.

4.6.2.3 Nike's Manufacturing Strategy and Stakeholder Implementation

(i) **Economic:** Nike, one of the world's leading retailers, sells a diverse range of garment goods, each needing unique production techniques and resources. Nike established a waste management facility to combine production materials to promote effective resource usage. This centre assists manufacturers in identifying waste sources and developing solutions, such as cross-functional materials efficiency teams, to decrease production waste and maximize resource utilization.

Using various technologies and programmes, Nike tackles essential environmental aspects of production, such as water pollution, carbon emissions, and waste. The business created a 'Water Risk Mitigation Guideline' to help manufacturers manage water use and treatment equipment. Nike also unveiled 'Colour dry' technology in 2012, which recycles CO_2 to substitute water in the dyeing process. Furthermore, Nike creates efficient tools and cutting patterns to reduce energy usage and waste during production (Gonçalves & Silva, 2021).

(ii) **Social:** Nike emphasizes worker participation in labour-intensive production. It performed social welfare trials involving over 28,000 workers in Vietnam, Indonesia, China, and Thailand, creating a protocol standard for measuring employee engagement. This resulted in lower turnover and absenteeism, suggesting increased worker well-being and engagement.

4.6.2.4 Nike's Distribution and Consumption Strategy and Stakeholder Implementation

(i) **Economic:** To react to the changing retail landscape, Nike established the 'NIKE Direct initiative' in 2017 to provide tailored service on a large scale. This initiative uses a variety of retail channels and technology, including mobile applications, to serve individual clients according to their preferences.

Nike also purchased Zodiac, a data analytics start-up, to better understand its consumers and provide specialized goods and services.

(ii) **Environmental:** Nike launched the Supplier Chain Sustainability Index (SCSI) to reduce carbon emissions from their supplier chains. Logistics partners must adhere to sustainability standards, and delivery solutions that emit fewer carbon emissions, such as bike delivery, are discovered and implemented. Nike has installed energy management systems and updated lighting in its stores to save energy.

(iii) **Social:** Through initiatives like the 'BIKETOWN' programme, which sets up bike stations across Portland, Nike has promoted sustainable consumption by introducing the 'Reuse-A-Shoe' programme. Customers drop off athletic shoes at retail stores to recycle into Nike Grind, which is to be used for building tracks, courts, playgrounds, or fields.

4.6.2.5 Performance Evaluation of Nike's Sustainable Planning Strategy

(a) Nike's renewable energy use has steadily increased from 14% in 2015 to 22% in 2017. The company aims to achieve 100% recyclable energy by 2020.

(b) The initiative for minimum water consumption resulted in a 7.4% reduction in freshwater use by 2017.

(c) Nike trained 82% and 72% of its manufacturers and suppliers on the AFIRM Group's restricted substance list (RSL), respectively, with 98% compliance.

(d) All factories received ratings higher than Bronze, meeting Nike's Sustainable Manufacturing and Sourcing Index (SMSI) guidelines.

(e) The 'U.S. Family Care Benefit Programme' debuted in 2016 and provides employees with paid time off for family care.

(f) Nike has engaged in community collaborations, such as with the Ministry of Education in China, where over 7000 students were educated to conduct sports courses and sports-related materials were delivered to over 300 schools.

4.6.2.6 Performance of Design Stakeholders

(a) Nike's Flyknit technology, which was introduced in 2015, increased sales and cut production waste by 60%.

(b) The footwear industry's revenue grew from $13.5 billion in the year 2012 to over $20 billion in 2018.

(c) In 2017, Nike employed eco-friendly rubber (98%) and its Flyknit core yarns, which were entirely made of recycled polyester.

(d) Nike converted more than 4.6 billion throwaway bottles into sportswear, demonstrating their dedication to ecologically beneficial products.

4.6.2.7 Performance of Manufacturing Stakeholders

(a) Nike's revenue grew 8% in 2017 thanks to increasing sales and streamlined manufacturing methods.
(b) Nike footwear completed items now use 60% less energy than 10 years ago.
(c) The 'Water Risk Mitigation Guideline' and 'Colour Dry' technology saved 20 million litres of water and cut waste by 60% during the midsole moulding process.

4.6.2.8 Performance of Distribution and Consumption Stakeholders

(a) Nike's 'direct to customer' retailing approach, 'Nike Direct,' resulted in a rise in global sales, including an 18% increase in digital direct sales.
(b) The Energy Management System (EMS) managed power usage in retail, resulting in only a 0.5% rise in energy consumption despite an 18% increase in sales.

4.6.2.9 Overall Performance of Nike's Sustainable Fashion Supply Chain

(a) Nike's sustainability initiatives encompass economic, ecological, and social dimensions, and various programmes and regulations are adopted across the supply chain (Fulton & Lee, 2013).
(b) Despite progress, there is always potential for development, notably in logistics and stakeholder involvement, to achieve better sustainability results (Fung et al., 2019).

4.6.3 Strategies for Small- and Medium-Sized Retailers to Adopt Sustainable Practices

Small- and medium-sized businesses (SMEs) are accomplishing quick development and disruptive innovations by focusing on entrepreneurial sustainability, which is uncommon in giant multinational corporations. This difference can be related to the very nature of sustainability. Hart and Milstein developed a strategy framework to help businesses examine investments through the' sustainable enterprise' lens to maximize profitability while promoting sustainability.

The framework assists enterprises in addressing conflicts caused by global factors such as rising human population, environmental constraints, energy and raw material supply, globalization, and social challenges. Larger companies frequently focus on incremental improvements in current operations, such as pollution avoidance and sustainable efficiency, but SMEs might compete more effectively by focusing on disruptive prospects.

SMEs bring skill sets to bear on disruptive prospects swiftly and competitively because they focus on sustainability from the start, without the distraction of fixing current operations. These SMEs may invest in cutting-edge initiatives without being constrained by established compliance-only investment and infrastructure. They have succeeded in various industries, including health care, communications, fashion, and manufacturing, by incorporating social and environmental sustainability into financial predictions and strategic corporate objectives (Todeschini et al., 2017).

The issue remains whether larger corporations will find it more cost-effective to join sustainable markets by investing in or acquiring SMEs or developing internal competencies. The question remains whether larger firms will find it more cost-effective to enter sustainable markets through investment, acquiring SMEs, or building internal capabilities. Organizational challenges within larger organizations might occur when new units compete with existing activity. Small and medium-sized enterprises (SMEs) can flourish without such competitive friction and expectations, allowing for faster technological and market progress.

Networked SMEs provide the financial and organizational efficiency required for long-term development, allowing the creation of breakthrough innovations and markets. Networking may reduce disruptive impacts from individual SMEs and accelerate progress towards sustainable growth, but it requires collaboration and solid business ecosystems.

Developing networked SME systems for sustainable supply chain management can help MNEs improve their economic performance in overseas markets (Moore & Manring, 2009). As creative destruction intensifies in the global marketplace, SMEs display the potential to flourish in new places, implying that new networked organizational structures may develop, with MNEs providing infrastructure and money and SMEs driving sustainable innovation. Collaboration will be critical for solving systemic issues that underpin industrial ecology, corporate resilience, and global sustainability (Oelze et al., 2020; Su et al., 2021).

4.7 Conclusion

Sustainable clothing retailers can overcome challenges by embracing innovative sourcing, strategic partnerships, and blockchain technology, which enhances traceability, transparency, and overall sustainability in the apparel industry.

Blockchain technology revolutionizes textile and apparel supply chains by boosting traceability, preventing fraud, and ensuring sustainability. Transparency and decentralized records ultimately foster greater consumer trust.

Integrating sustainability into fashion design requires overcoming knowledge gaps, material constraints, and internal obstacles. However, successful integration also requires collaboration between the fashion industry and informed consumers, which is crucial in driving and shaping these changes.

Adopting eco-friendly and green packaging meets regulatory demands and consumer preferences, ultimately lowering costs, boosting brand reputation, and advancing sustainability. This practice benefits both environmental and economic performance.

Nike's comprehensive manufacturing and stakeholder strategies address economic, environmental, and social impacts. They reveal significant advancements in resource efficiency, waste reduction, and worker well-being while revealing areas for further improvement.

4.8 Future Trends and Outlook

The future of the garment and textile industry is typically discussed in terms of national trade flows. These flows, however, are moulded by the actions of private players throughout supply chains—retailers and producers—who operate within the context of national and international commerce and other state policies. Understanding how these private decisions influence observed garment movements is critical for projecting the changing post-2005 market (Abernathy et al., 2006).

The circular economy concept, which highlights reuse, reduction, and recycling, influences production and consumption patterns, impacting how things are manufactured, distributed, and consumed. Studies on green garments reveal three major themes connected to the circular economy: supply chain difficulties, business model adjustments, and designer roles. Supply chain sustainability is a primary priority, with measures such as partnering with retailers, improving material environmental quality, and investing in cost-effective solutions for small supply businesses. Business model innovation is critical, necessitating changes in ideation, product development, production, retailing techniques, repair, reuse, and recycling. Designers play a pivotal role in influencing material choices, promoting awareness, encouraging waste reduction, and supporting sustainable practices, while some designers are known to be unfamiliar with eco-friendly materials (Urmínová & Kusá, 2021; Mizrachi & Tal, 2022; Ray & Nayak, 2023).

Consumer knowledge is critical in shaping green apparel purchasing behaviour. Learning programmes and formal instruction can positively impact pro-environmental sentiments. However, hurdles like the cost, availability, and benefits of eco-friendly clothing might deter real purchasing habits. Advertising efforts concentrating on environmental problems and celebrity endorsements can affect customer behaviour, albeit their impact may be greater on opinions than actual behaviours. The media, particularly social media, can potentially influence customer attitudes and actions around green clothes (Harris et al., 2016; Hedegård et al., 2019).

Several challenges exist to the mainstream adoption of green garments, including consumption patterns, a lack of availability, high prices, transparency concerns, and views of green gear as unfashionable. Overcoming these hurdles involves addressing customer attitudes and behaviours through education, effective advertising, and changes in company structures and supply chain policies (Lang et al., 2016; Mukendi et al., 2020; Khan et al., 2023).

References

Abbate, S., Centobelli, P., Cerchione, R., Nadeem, S. P., & Riccio, E. (2024). Sustainability trends and gaps in the textile, apparel and fashion industries. *Environment, Development and Sustainability, 26*(2), 2837–2864). Springer Science and Business Media B.V. https://doi.org/10.1007/s10668-022-02887-2

Abernathy, F. H., Volpe, A., & Weil, D. (2006). The future of the apparel and textile industries: Prospects and choices for public and private actors. *Environment and Planning A, 38*(12), 2207–2232. https://doi.org/10.1068/a38114

Ahmed, W. A. H., & Maccarthy, B. L. (2021). Blockchain-enabled supply chain traceability in the textile and apparel supply chain: A case study of the fiber producer, lenzing. *Sustainability (Switzerland), 13*(19), 10496. https://doi.org/10.3390/su131910496

Ara, H., Leen, J. Y. A., & Hassan, S. H. (2019). GMS for sustainability performance in the apparel manufacturing industry: A conceptual framework. *Vision, 23*(2), 170–179. https://doi.org/10.1177/0972262919850931

Baker, K., & McNeill, L. S. (2023). A five-stage framework for sustainability certification for fashion brands: Can a standardised approach work in the fashion industry? *International Journal of Fashion Design, Technology and Education, 17*, 287. https://doi.org/10.1080/17543266.2023.2277263

Bhandari, N., Garza-Reyes, J. A., Rocha-Lona, L., Kumar, A., Naz, F., & Joshi, R. (2022). Barriers to sustainable sourcing in the apparel and fashion luxury industry. *Sustainable Production and Consumption, 31*, 220–235. https://doi.org/10.1016/j.spc.2022.02.007

Blissick, M., Dickson, M. A., Silverman, J., & Cao, H. (2017). Retailers' extent of involvement in sustainability and role in creating sustainable apparel and textiles from South Africa. *International Journal of Fashion Design, Technology and Education, 10*(3), 265–275. https://doi.org/10.1080/17543266.2017.1298160

Caniato, F., Caridi, M., Crippa, L., & Moretto, A. (2012). Environmental sustainability in fashion supply chains: An exploratory case-based research. *International Journal of Production Economics, 135*(2), 659–670. https://doi.org/10.1016/j.ijpe.2011.06.001

Curwen, L. G., Park, J., & Sarkar, A. K. (2013). Challenges and solutions of sustainable apparel product development: A case study of Eileen Fisher. *Clothing and Textiles Research Journal, 31*(1), 32–47. https://doi.org/10.1177/0887302X12472724

Dabija, D. C., & Băbut, R. (2019). Enhancing apparel store patronage through retailers' attributes and sustainability. A generational approach. *Sustainability (Switzerland), 11*(17). https://doi.org/10.3390/su11174532

Dal Mas, F., Tucker, W., Massaro, M., & Bagnoli, C. (2022). Corporate social responsibility in the retail business: A case study. *Corporate Social Responsibility and Environmental Management, 29*(1), 223–232. https://doi.org/10.1002/csr.2198

ElMessiry, M., & ElMessiry, A. (2018). Blockchain framework for textile supply chain management: Improving transparency, traceability, and quality. *Lecture notes in computer science (Including subseries lecture notes in artificial intelligence and lecture notes in bioinformatics), 10974 LNCS*, 213–227. https://doi.org/10.1007/978-3-319-94478-4_15.

Fratto, G. M., Jones, M. R., & Cassill, N. L. (2006). An investigation of competitive pricing among apparel retailers and brands. *Journal of Fashion Marketing and Management: An International Journal, 10*(4), 387–404. https://doi.org/10.1108/13612020610701938

Fulton, K., & Lee, S.-E. (2012). An exploratory study of sustainable apparel retailers on the internet. *In Int. J. Electronic Marketing and Retailing, 5*(2), 95–109. https://doi.org/10.1504/IJEMR.2012.051027

Fulton, K., & Lee, S. E. (2013). Assessing sustainable initiatives of apparel retailers on the internet. *Journal of Fashion Marketing and Management, 17*(3), 353–366. https://doi.org/10.1108/JFMM-11-2012-0071

Fung, Y. N., Choi, T. M., & Liu, R. (2019). Sustainable planning strategies in supply chain systems: Proposal and applications with a real case study in fashion. *Production Planning and Control, 31*(11–12), 883–902. https://doi.org/10.1080/09537287.2019.1695913

Gam, H. J., & Banning, J. (2011). Addressing sustainable apparel design challenges with problem-based learning. *Clothing and Textiles Research Journal, 29*(3), 202–215. https://doi.org/10.1177/0887302X11414874

Gonçalves, A., & Silva, C. (2021). Looking for sustainability scoring in apparel: A review on environmental footprint, social impacts and transparency. *Energies, 14*(11), 3032. MDPI AG. https://doi.org/10.3390/en14113032

Goworek, H. (2011). Social and environmental sustainability in the clothing industry: A case study of a fair trade retailer. *Social Responsibility Journal, 7*(1), 74–86. https://doi.org/10.1108/17471111111114558

Goworek, H., Fisher, T., Cooper, T., Woodward, S., & Hiller, A. (2012). The sustainable clothing market: An evaluation of potential strategies for UK retailers. *International Journal of Retail and Distribution Management, 40*(12), 935–955. https://doi.org/10.1108/09590551211274937

Hagan, J., & Sporn, L. (2018). *Sustainability strategies and challenges in the luxury apparel industry*. Thesis for: Master of Environmental Studies. https://doi.org/10.13140/RG.2.2.27380.19844.

Harris, F., Roby, H., & Dibb, S. (2016). Sustainable clothing: Challenges, barriers and interventions for encouraging more sustainable consumer behaviour. *International Journal of Consumer Studies, 40*(3), 309–318. https://doi.org/10.1111/ijcs.12257

Hedegård, L., Gustafsson, E., & Paras, M. K. (2019). Management of sustainable fashion retail based on reuse– A struggle with multiple logics. *International Review of Retail, Distribution and Consumer Research, 30*(3), 311–330. https://doi.org/10.1080/09593969.2019.1667855

Helmore, E. (2018). *Nike hit with lawsuit from four women who allege gender discrimination*. Accessed on 13 Sept 2024. https://www.theguardian.com/business/2018/aug/10/nike-lawsuit-women-gender-discrimination

Hepburn, S. (2015). *Nike and Adidas show cautious support for eco-friendly dye technology*. Accessed on 13 Sept 2024. https://www.theguardian.com/sustainable-business/sustainable-fashion-blog/2015/apr/24/nike-and-adidas-show-cautious-support-for-eco-friendly-dye-technology

Islam, M. M., Perry, P., & Gill, S. (2021). Mapping environmentally sustainable practices in textiles, apparel and fashion industries: A systematic literature review. *Journal of Fashion Marketing and Management, 25*(2), 331–353. https://doi.org/10.1108/JFMM-07-2020-0130

James, F., & Kurian, A. (2021). Sustainable packaging: A study on consumer perception on sustainable packaging options in e-commerce industry. *Natural Volatiles and Essential Oils, 8*(5), 10547–10559.

Jung, S., & Jin, B. (2014). A theoretical investigation of slow fashion: Sustainable future of the apparel industry. *International Journal of Consumer Studies, 38*(5), 510–519. https://doi.org/10.1111/ijcs.12127

Jung, S., Jeong, C. H., & Ellie Jin, B. (2020). Does effective cost transparency increase price fairness? An analysis of apparel brand strategies. *Journal of Brand Management, 27*, 495–507. https://doi.org/10.1057/s41262-020-00191-w

Kang, Jiyun, and Gwendolyn Hustvedt (2014). "Building trust between consumers and corporations: The role of consumer perceptions of transparency and social responsibility." *Journal of business ethics* 125: 253–265.

Khan, S. J., Badghish, S., Kaur, P., Sharma, R., & Dhir, A. (2023). What motivates the purchasing of green apparel products? A systematic review and future research agenda. *Business Strategy and the Environment, 32*(7), 4183–4201. https://doi.org/10.1002/bse.3360

Köksal, D., Strähle, J., & Müller, M. (2018). Social sustainability in apparel supply chains—The role of the sourcing intermediary in a developing country. *Sustainability (Switzerland), 10*(4). https://doi.org/10.3390/su10041039

Lamrad, N. (2013). Manufacturing governance: Global norms and policy diffusion – The case of the Chinese apparel industry. *Journal of Asian Public Policy, 6*(2), 147–162. https://doi.org/10.1080/17516234.2013.814307

Lang, C., Armstrong, C. M., & Liu, C. (2016). Creativity and sustainable apparel retail models: Does consumers' tendency for creative choice counter-conformity matter in sustainability? *Fashion and Textiles, 3*(1).

Leonora, F., Iris, M., Ali, B. M., & Nicholas, G. (2022). The new 3Ps of sustainability marketing: The case of fashion. *Sustainable Production and Consumption, 31*, 384–396. https://doi.org/10.1016/j.spc.2022.03.004

Mann, M., & Byun, S. E. (2011). Accessing opportunities in apparel retail sectors in India: Porter's diamond approach. In. *Journal of Fashion Marketing and Management, 15*(2), 194–210. https://doi.org/10.1108/13612021111132636

McCormick, H., Cartwright, J., Perry, P., Barnes, L., Lynch, S., & Ball, G. (2014). Fashion retailing – Past, present and future. *Textile Progress, 46*(3), 227–321. https://doi.org/10.1080/00405167.2014.973247

Mizrachi, M. P., & Tal, A. (2022). Regulation for promoting sustainable, fair and circular fashion. *Sustainability (Switzerland), 14*(1). https://doi.org/10.3390/su14010502

Moon, K. K. L., Youn, C., Chang, J. M. T., & Yeung, A. W. H. (2013). Product design scenarios for energy saving: A case study of fashion apparel. *International Journal of Production Economics, 146*(2), 392–401. https://doi.org/10.1016/j.ijpe.2013.02.024

Moore, S. B., & Manring, S. L. (2009). Strategy development in small and medium sized enterprises for sustainability and increased value creation. *Journal of Cleaner Production, 17*(2), 276–282. https://doi.org/10.1016/j.jclepro.2008.06.004

Mukendi, A., Davies, I., Glozer, S., & McDonagh, P. (2020). Sustainable fashion: Current and future research directions. *European Journal of Marketing, 54*(11), 2873–2909). Emerald Group Holdings Ltd. https://doi.org/10.1108/EJM-02-2019-0132

Naidoo, M., & Gasparatos, A. (2018). Corporate environmental sustainability in the retail sector: Drivers, strategies and performance measurement. *Journal of Cleaner Production, 203*, 125–142). Elsevier Ltd. https://doi.org/10.1016/j.jclepro.2018.08.253

Nayak, R. (2019). A review of recent trends in sustainable fashion and textile production. *Current Trends in Fashion Technology & Textile Engineering, 4*(5). https://doi.org/10.19080/ctftte.2019.04.555648

Oelze, N., Gruchmann, T., & Brandenburg, M. (2020). Motivating factors for implementing apparel certification schemes—A sustainable supply chain management perspective. *Sustainability (Switzerland), 12*(12). https://doi.org/10.3390/SU12124823

Pal, R. (2014). *Sustainable business development through designing approaches for fashion value chains* (pp. 227–261). https://doi.org/10.1007/978-981-287-110-7_9.

Ray, S., & Nayak, L. (2023). Marketing sustainable fashion: Trends and future directions. *Sustainability (Switzerland), 15*(7), 6202). MDPI. https://doi.org/10.3390/su15076202

Rogetzer, P., Silbermayr, L., & Jammernegg, W. (2018). Sustainable sourcing of strategic raw materials by integrating recycled materials. *Flexible Services and Manufacturing Journal, 30*(3), 421–451. https://doi.org/10.1007/s10696-017-9288-4

Sanmiguel, P., Pérez-bou, S., Sádaba, T., & Mir-bernal, P. (2021). How to communicate sustainability: From the corporate web to E-commerce. The case of the fashion industry. *Sustainability (Switzerland), 13*(20). https://doi.org/10.3390/su132011363

Shen, B., Li, Q., Dong, C., & Perry, P. (2017). Sustainability issues in textile and apparel supply chains. *Sustainability (Switzerland), 9*(9). https://doi.org/10.3390/su9091592

Su, J. (2013). Strategic sourcing in the textile and apparel industry. *Industrial Management and Data Systems, 113*(1), 23–38. https://doi.org/10.1108/02635571311289647

Su, J., Wood, A. M., & Gargeya, V. B. (2021). Sustainable entrepreneurship in the apparel industry: Passion and challenges. *Journal of the Textile Institute, 113*(9), 1935–1941. https://doi.org/10.1080/00405000.2021.1957276

Sun, O. (2018). *Sustainability strategies and challenges in the luxury apparel industry*. https://doi.org/10.13140/RG.2.2.27380.19844.

Todeschini, B. V., Cortimiglia, M. N., Callegaro-de-Menezes, D., & Ghezzi, A. (2017). Innovative and sustainable business models in the fashion industry: Entrepreneurial drivers, opportunities, and challenges. *Business Horizons, 60*(6), 759–770. https://doi.org/10.1016/j.bushor.2017.07.003

Urmínová, M., & Kusá, A. (2021). Innovative approaches in marketing communication in sustainable fashion business. In *Proceedings of the European Conference on Innovation and Entrepreneurship, ECIE, 2020-September*, pp. 834–841. https://doi.org/10.34190/EIE.20.183

Vadakkepatt, G. G., Winterich, K. P., Mittal, V., Zinn, W., Beitelspacher, L., Aloysius, J., Ginger, J., & Reilman, J. (2021). Sustainable Retailing. *Journal of Retailing, 97*(1), 62–80. https://doi.org/10.1016/j.jretai.2020.10.008

Vadicherla, T., & Saravanan, D. (2015). *Sustainable measures taken by brands, retailers, and manufacturers* (pp. 109–135). Springer. https://doi.org/10.1007/978-981-287-164-0_5

Warasthe, R., Schulz, F., Enneking, R., & Brandenburg, M. (2020). Sustainability prerequisites and practices in textile and apparel supply chains. *Sustainability (Switzerland), 12*(23), 1–19. https://doi.org/10.3390/su12239960

Yang, S., Song, Y., & Tong, S. (2017). Sustainable retailing in the fashion industry: A systematic literature review. *Sustainability (Switzerland), 9*(7), 1266). MDPI. https://doi.org/10.3390/su9071266

Chapter 5
Circular Packaging Models in Apparel Retailing

R. Surjit, B. Jenella, K. Thejesvikashni, V. Prabalaya, and L. Sasidevi

5.1 Introduction

Packaging used in a product has significant importance concerning environmental impact if it is not sustainable and uses a lot of artificial elements that cause pollution. The product packaging is an important marketing element that boosts the sales of a product as customers look for quality, comfort, and aesthetic packing. There is a great need for manufacturers and brands to pack their products as required by the customer so that they are satisfied with the product and packaging. Due to the increased customer-centric approach, brands or manufacturers miss the sustainability aspect of packaging and use elements in packing that are not sustainable and cause harm to the environment and society. The packaging is expected to be safe till it reaches the customer and for that, many synthetic items are used which cause problems to the environment in the long run. Many industries overuse packaging resources, leading to environmental damage for many years now. In many countries, there is a strict regulation to follow the rules to reduce or eliminate packaging further, which helps to decrease dependency on packaging. Packaging materials become a waste product once the product that was packaged in it is opened or taken from it for the intended use, but effective packaging can reduce the wastage of items. The cause of wastage in packaging is the large number of materials used by the packaging industries to fulfil a particular function. By using the latest eco-friendly materials, and improving manufacturing and design of packaging materials, environmental conservation can be improved without compromising the quality

R. Surjit (✉) · B. Jenella · K. Thejesvikashni · V. Prabalaya · L. Sasidevi
Department of Fashion Technology, PSG College of Technology,
Coimbatore, Tamil Nadu, India
e-mail: rst.fashion@psgtech.ac.in; 22H213@psgtech.ac.in; 22H241@psgtech.ac.in;
22H249@psgtech.ac.in; 22H251@psgtech.ac.in

S. S. Muthu (ed.), *Sustainable Apparel Retailing*, Sustainable Textiles: Production, Processing, Manufacturing & Chemistry,
https://doi.org/10.1007/978-3-031-83875-0_5

of materials. In the case of packaging materials also, 4 Rs can be used which are 'Reduce, Reuse, Recycle, and Redesign'. The packing materials can be reduced effectively or reused or recycled or redesigned to meet the needs. Packaging also aims to ensure safety, enhance support, and efficient transportation. In packaging, manufacturers deliver products to consumers in pristine condition and create a visual function experience. In the case of apparel packaging, the motive is to protect or secure garments from damage and make them presentable to customers. Packaging and package are business functions and items, respectively. In many situations, packaging is more important than the product it contains, which involves information about the product like labels, inserts, how to use it, and sizes of packaging of the products.

Packaging material also causes pollution and questions sustainability as there is a huge amount of waste generated because of packaging materials after usage. According to the Central Pollution Control Board (CPCB), India generates 26,000 tonnes of plastic waste daily of which 43% of plastic waste is from the packaging industry (Dhiman et al., 2022). There is an impact on air pollution because of transportation as packaging of products adds weight to the product which leads to an increase in fuel consumption and emission of carbon dioxide. It leads to land and water contamination as improper management of packaging waste may lead to soil and water body contamination like landfills, and pollutants in groundwater. Also, the production of packaging materials needs the extraction of resources like timber, oil, and minerals which may lead to deforestation and soil degradation. So, it is imperative to understand the basics of packaging and how it can be improved to make it sustainable. This chapter details the functions of packaging, types of packaging materials, apparel packaging types, employing 7Rs for improving sustainability in packaging, initiatives taken by apparel brands in using eco-friendly packing materials, and the various tools used to measure sustainable practises in packaging.

5.1.1 Functions of Packaging

Packaging is done for product protection, product preservation, improving marketing and branding, improving convenience and user experience, and also involves sustainable and eco-conscious packaging. The two main functions of apparel packaging involve distribution function and merchandising function. The distribution function deals with the packaging of apparel products in a way that helps the manufacturer to send it to the retailer or the wholesaler at the lowest cost and shortest time. The merchandising function deals with the packaging in a manner that will stimulate the customer to buy the product or the final packaging that the customer will get when he or she buys the product. However, both functions should ensure that the quality, durability, and style specifications are intact till the material reaches the customer.

Product Protection

Packaging plays a major role in protecting products from various external factors such as impact, vibration, temperature, moisture, and tampering. Techniques like cushioning materials, shock-absorbing components, moisture-resistant coatings, and tamper-evident seals contribute to the product's safety and integrity throughout its journey from manufacturer to consumer. Productive packaging of a product is important because when the product is shipped, the customer demands the product should be safe and secure and arrives with no damage. The materials that can be filled in the protective packaging of the product are paper fill, bubble packaging, air pillow, foam packaging, etc. Similarly, apparel packaging should be done in such a way that the product inside the package has no defects or issues generated because of packaging.

Product Preservation

Packaging aids in preserving product freshness, quality, and shelf life. Techniques like vacuum sealing, modified atmosphere packaging (MAP), and intelligent packaging systems help maintain optimal conditions such as controlled oxygen levels and humidity, thereby extending the product's viability. Packaging makes it easy to store and usage of products. It may be so designed to make it easy to handle by the consumers through the packaging and also easy for the distributors.

Marketing and Branding

Packaging is a powerful tool in marketing to attract consumers and promote brand identity. To approach consumers, visual design elements, logos, colour schemes and details about the product on packaging help to differentiate products from competitors. Packaging design can evoke positive associations, influencing consumers purchasing decisions. Packaging is also as important as the product, which is important for product brand and marketing. The uniqueness of packaging can increase the willingness to buy the product. The packaging of a product can tell the story about the company's value of producing the product and considering the social and environmental impact, which brings benefit to the customers for buying products.

Convenience and User Experience

Apparel product packaging design gives the satisfaction of convenience and improves user experience to the consumers which helps them to access the products. A positive user experience contributes to consumer satisfaction and loyalty. Nowadays, consumers are more concise about the packaging of user-friendly products and life-saving materials. Packaging should be easy to open and close and easy to fold. More precisely, it should be reusable and recyclable to satisfy consumer's needs. People are bringing back the trends of before the 1960s of single-use culture to buy in bulk with their containers, which is more beneficial for environmentally friendly purchasing.

Sustainability and Eco-Conscious Packaging
The importance of Sustainable Packaging cannot be overstated. It is to be understood that eco-conscious circular packaging is the need of the hour. Circularity is a part of sustainability and is encompassed by it. Producing circular packaging materials will in turn lead to the sustainability of the environment as circularity is one of the ways to achieve sustainability and reduce the environmental impact. For example, sustainability can be achieved in various ways like utilising fewer raw materials, eco-friendly materials, and circular business models. In these ways, if in case, more raw materials are used out of necessity, then it may lead to environmental impact but by using circular packaging methods like 7R, the materials can still be reused, repurposed, recycled and the time for the packaging material to reach the grave (end of life, thrown as waste) can be reduced. So circular packaging is essential. The packaging industry is increasingly exploring eco-friendly alternatives, including plant-based bio-plastics, recycled materials, and compostable packaging. By mitigating environmental impact, optimising materials usage, promoting recycling and exploring innovative solutions, packaging can contribute to a more sustainable future.

5.1.2 Classification of Packaging Materials

Paper-Based Packaging
Paper-based packaging includes paperboard and cardboard, each with its applications and characteristics. Paperboard is thicker and stiffer than regular paper and is commonly used to create boxes and cartons. Cardboard refers to a heavier and more durable form of paper-based packaging material. Paper-based packaging has several benefits beyond its utility. It is often sourced from renewable materials and is biodegradable, making it a more sustainable choice compared to other packaging options. However, it is crucial to balance the environmental impact with the need for protection and durability.

Plastic Packaging
Plastic packaging has become ubiquitous due to its versatility, durability, and cost-effectiveness. Common plastic materials used in packaging include polyethylene, polypropylene, and polyethylene terephthalate (PET). These plastic materials can be moulded, extruded, or formed into a variety of shapes and sizes to suit different packaging requirements. While plastic packaging offers exceptional protection and convenience, it also presents challenges in terms of environmental impact. Plastic waste has become a significant concern, leading to initiatives aimed at reducing plastic consumption and encouraging recycling. However, advancements in eco-friendly plastics, such as biodegradable and compostable alternatives, provide hope for a more sustainable plastic packaging industry.

Glass Packaging
Glass packaging has a long-standing history and is revered for its unique properties. This type of material provides excellent barrier properties, protecting the contents from contamination, while also being impermeable to gases and moisture. Additionally, glass has an aesthetic appeal that can enhance the perception and desirability of the product it contains. It has extensive usage in industries such as **beverages**, cosmetics, and **pharmaceuticals**. Its transparency allows consumers to view the product, instilling confidence and establishing a sense of authenticity. However, glass packaging is relatively fragile and requires careful handling during transportation and storage.

Metal Packaging
Metal packaging, commonly made from aluminium or steel, offers unparalleled strength and durability. It is extensively used for canned goods, aerosol containers, and beverage cans. Metal material provides an excellent barrier against oxygen, light, and moisture, protecting products from spoilage and extending their shelf life.

Other Packaging Materials
Metal packaging, commonly made from aluminium or steel, offers unparalleled strength and durability. It is extensively used for canned goods, aerosol containers, and beverage cans. Metal material provides an excellent barrier against oxygen, light, and moisture, protecting products from spoilage and extending their shelf life.

5.2 Apparel Packaging

In our day-to-day life, packaging plays a major role and often fulfils our needs, from day to night. We encounter various forms, from toothpaste tubes to leftovers stored in food containers, manufacturers have evolved significantly, adapting to changing needs, preferences, and technological advancements. Packaging is a word that is used to denote the art and science of packing (Gray, 1939). It is used for securing and safekeeping products for sales to customers or for effective transportation of goods between businesses.

The apparel packaging is done in two different ways. One is the customer packing (also known as merchandising packing) and the other one is the shipping packing. Shipping packaging is the packaging done for finished apparel products to send to the retailer for selling and customer packing is the packing done for delivering the products to the customer. In the case of shipping packaging, the apparel is sometimes kept in hangers and hung on hanger stands as open packing and transported in containers for the retailer to showcase it as it is in the store. Otherwise, the customer packing (merchandise packing) is done and then put in carton boxes for transporting to the retailer. The customer packaging involves various packaging materials as shipping packaging can be done with carton boxes or hangers. The merchandising packaging uses polythene bags, transparent covers and paper shirt boxes and then they are put in carton boxes for shipping packaging. To give identification or information to the customers other packaging materials are also attached to the apparel products. These include hang tags, header cards, size labels, wash care labels, barcode stickers, etc. In many cases, especially for online retailing, the product is packed as per the requirement of the buyer's choice which may be any type of packing. The apparel packaging process involves the process of tagging the product, then folding it, followed by poly bag packing and finally put in carton boxes and the cartons are strapped.

Since packaging plays crucial role in the apparel industry, there is a need to understand the difference between regular packaging and sustainable packaging. As known, the materials used in packaging such as plastics, glues, and tape are the materials that cannot be recycled and are non-biodegradable. Plastic packaging contributes to pollution during its production, disposal, and usage which is the emission of greenhouse gases and resource depletion. There is a need for sustainable packaging due to the following benefits.

1. *Environmental benefits*: Sustainable packaging decreases the use of renewable materials, reduces wastage and enhances the eco-friendly process. It allows the usage of biodegradable and compostable materials, which decompose easily and protects the environment.
2. *Financial benefits*: It leads to cost savings in the long run since it reduces packaging waste so the disposal cost is reduced. The energy and resources used for it are also less.
3. *ESG compliance*: Environmental, social, and governance (ESG) factors are significantly important to consumers and investors. By following sustainable packaging, apparel companies can showcase their commitment to contributing to a sustainable environment to enhance their brand name and publicity to consumers (Pollman, 2019).

5.2.1 Types of Apparel Packaging

The apparel packaging can be categorised into three main types:

Primary Packaging

The primary or first layer of packaging that comes in direct contact with the product. It is designed to provide both protection and functionality. Examples in general are bottles, cans, pouches, blister packs, and clamshell containers. Examples in apparel would be the transparent covers in which they are put. It ensures product integrity, prevents contamination, and can even contribute to the visual appeal.

Secondary Packaging

Secondary packaging or middle layer surrounds primary packaging, serving as a supplementary layer for product grouping, and protection during distribution, and transportation. Common examples are Boxes, trays, labels, and shrink wraps. It not only safeguards the primary packaging but also enables efficient handling, branding, and marketing efforts.

Tertiary Packaging
Tertiary packaging or the outermost layer is used for bulk transportation and storage. It typically involves packaging materials like pallets, containers, and stretch wraps. Tertiary packaging ensures the safety and movement of products within the supply chain and provides a foundation for logistics processes such as stacking, lifting, and securing goods.

Apart from these types, concerning apparel, there is shipment packaging and merchandising packaging as mentioned above. The merchandising packaging is the packaging in which the customer receives the product and it should be in such a way that the product can be identified, the product's appeal gets enhanced because of packaging, the customer gets attracted because of the packaging, and the product is safe till the customer starts using it. Also, from the retailer's point of view, such packages should increase customer appeal, handling cost of the package should be less, easy to store with fewer space requirements, and easy to handle with minimal time and effort for storing, moving it to selling area, and finally for handing it over to the customer. In the case of shipment packaging, it is the distribution function where the carrier receives and delivers the package to the retailer from the manufacturer. These packages should be in such a way that the garment quality is protected and preserved and handling costs are to be minimal. The shipping packaging can be put in closed containers where each garment is covered completely as in the case of merchandise packaging for delivery to the customer or without a covering for displaying in retail stores. They can also be transported in open containers for smaller distances in closed or open merchandising packages.

5.2.2 Packaging Materials

Garment packaging is the most important process in the production of garments which enables to distribution of the final product to the consumer. So the materials used in packaging are more important than any other systems in the industry. The major objectives of packaging are to prevent any damage to the product while transporting it and to enhance the overall ability of the product for the consumer sale for it.

Basic Materials Used in Packaging
Bags, boxes, cartons, cases, crates, twine, and wrappers are some of the basic materials used for packaging because of their thickness weight and strength properties to deliver the materials safely.

Types of Packaging Materials
The elementary materials used in packaging are paper, plastics, film, wood, nails, staples, cords gum, tape, and metal bands.

1. Wood cases and crates are used to export bulk materials or rugged shipments where the handling of the packed products is high.

2. Paper and plastic film packaging materials are widely used in apparel industries. The paper types such as kraft, crepe, and tissue are used. Paper foil, paper board, and waterproof materials are used in packaging materials. These plastic films are advantageous to paper because of their clarity.

5.2.3 Waste Generation and Lack of Sustainability

The packaging material generates a huge amount of waste. It is estimated that 77.7 million tons and 82.2 million tons of packaging waste were generated in Europe and the USA in 2018 (Jestratijevic et al., 2022). Also, plastic waste is so huge that if a single plastic bag is not properly disposed of, it may release greenhouse gases and pollute the environment for hundreds of years (Chamas et al., 2020). Such packaging wastes are a burden to the environment and cause detrimental effects to future generations. It is essential to work on mitigating the wastage generated by employing suitable methods. 7R can be an effective strategy for the same.

5.3 The 7Rs of Sustainable Packaging in the Apparel Industry

The apparel and fashion industry produces a large amount of packaging waste that challenges a sustainable environment. Packaging is mainly used to protect and transport textiles, apparel and footwear products to distribute those final finished goods to stores and consumers. So, packaging plays a huge part in the apparel industry which generates a lot of waste like plastic waste used to protect individual garments that cannot be recycled, cardboard waste, tags and labels, adhesive tapes, bubble wrap, etc. A report from Fashion Industry Waste Statistics in 2023 states that the fashion industry produces 97 million tons of waste per year out of which 18 million contribute leftover textiles, 2.5 million contribute chemical waste, and three million were packaging materials (Royal, 2024). These materials cause landfills, pollution, and emission of greenhouse gases. In concerning this, we can follow the "7Rs" framework in textile and apparel packaging not only to help to package in a more sustainable and eco-friendly manner but also to make the brand's sustainable credentials more efficient. The 7Rs include Reduce, Reuse, Recycle, Refuse, Repurpose, Rot and Rethink provide a comprehensive approach to addressing the environmental impact of packaging in this sector.

5.3.1 Reduce

The most essential step in following sustainability is to reduce waste. Consumers are very interested in supporting and buying products from brands that align with their beliefs and values. When the packaging of a particular brand is not sustainable, customers will prefer to purchase from their rival brand which does offer eco-friendly packaging. However, one cannot suddenly implement the purchase and preaching of sustainable things in companies but we can make small changes, like reducing the usage of non-biodegradable things, reducing the non-ecofriendly things used for packaging and encouraging recycling within the workplace. Product packaging waste can be reduced in apparel industries by encouraging employees to reduce the wastage. They can be taught best practises for handling packaging materials so that the materials do not get damaged during the packing process. The employees should be aware of the company's new methodology of reducing wastage of materials. It is important that the packaging has a good look but it should also keep our apparel product safe and environment friendly, some of the materials used in our packaging might be unnecessary. Minimising the packaging materials does not mean that the design aspects should be compromised. Simple packaging with visually appealing designs and protection can also be made by reducing extra unneeded materials. This will help the planet and also decrease production costs as fewer materials will be needed to purchase less energy will. Using eco-friendly alternatives will make packaging more sustainable. The usage of plastics can be reduced in packaging and other sustainable materials like bagasse, cork, PLA, hemp, recycled paper, and corrugated cardboard can be used for packing. More emphasis can be given to the development and usage of alternate packaging materials other than plastic so that they will not cause damage to the environment.

5.3.2 Reuse

The concept of reusability in packaging has played an important role in sustainability. Several businesses are taking step-forward initiatives to produce packaging that consumers can extend its life by recycling and reusing which eliminates the need for single-use alternatives. Brands should evaluate their packaging strategies and should make sure that materials can be reused at least once. For example, sturdy boxes can be used as storage bins and fabric pouches can be reused as reusable bags for small objects. It should also be noted that the packaging item which can be reused should also be sustainable. Products like bio-plastic bags, plantable bags, and plantable tags infused with seeds are some of the great technological innovations that can be used to lead to sustainable and eco-conscious purchases for consumers. The reusable packaging appears to be a powerful solution for the environmental concerns that arise from using single-use packaging. Since it is durable and harmless to the environment, it can benefit the brands by minimising waste generation and contributing more to a circular economy. The difference between reusable and single-use packaging is its lifecycle and function. Packaging used for single use is made to be used just once before being eliminated or recycled. Reusable packaging, on the other hand, can be refilled, returned or repurposed again. So single-packing materials that are disposed of after one-time use can cause landfills which leads to pollution meanwhile using reusable packaging reduces the need for continuous manufacture and elimination of packaging materials.

The widespread use of reusable packaging across various industries shows its adaptability and efficiency. Some examples that how apparel industries can implement reusability in packaging are given below.

Reusable Totes and Shopping Bags

Brands should suggest customers switch from single-use plastic bags to reusable totes or shopping bags produced from sturdy materials like recycled materials and cloth. These eco-friendly reusable bags provide an opportunity for businesses to follow a circular economy.

Reusable Shipping Containers

Reusable shipping containers are being used by logistics organisations and e-commerce businesses to minimise packaging waste. The requirement of extra cardboard boxes and packing materials is reduced as these are made from durable materials like metal or plastic which can be reused again and over again for shipments.

Return Packaging Initiatives

Some of the companies have come up with packaging return programmes that allow customers to return used packaging such as padded mailers or boxes, to specific drop-off locations or via a reverse logistics system. By inspecting, cleaning, and reusing these packaging materials, wastes can be reduced and circularity is promoted.

As the requirement for sustainable packaging solutions rises widely, more creative solutions are emerging to boost the efficiency and impact of reusable packaging. So, here are some methods to enhance reusable packaging.

Modular and Customised Solutions

Companies can choose reusable packaging solutions that are customised to meet the needs of their supply chain. Modular packaging solutions are versatile and support a variety of product sizes and configurations. These personalised packaging solutions guarantee effective handling and storage by reducing waste and increasing space usage throughout the supply chain.

Smart Reusable Packaging

Incorporating smart technology, such as RFID tags or sensors can completely change reusable packaging solutions. With the help of these technologies, packing assets may be tracked and monitored in real time, ensuring their proper administration, maintenance, and return. It also improves transparency which enables businesses to optimise logistics and pinpoint areas in need of improvement.

Combining Packaging Platforms
Platforms for collaboration are emerging to allow businesses to share and pooling of reusable packaging assets. These platforms facilitate the sharing and exchange of packing resources within the business working in the same sector or supply chain. Promoting collaboration and resource sharing can lower expenses, improve asset consumption, and increase the overall efficiency of reusable packaging systems.

Innovative Materials and Design
Discovering new materials and design ideas creates chances for more sustainable and beneficial reusable packaging. For example, biodegradable or compostable materials and bio-based plastics can be substituted as environmentally friendly alternatives to traditional packaging. Furthermore, creative designs that prioritise easy cleaning, durability, and simplicity with automated systems can increase the efficiency of reusable packaging.

Consumer Engagement and Education
Initiatives intended for promoting reusable packaging must be successful in boosting customer participation and engagement. Brands should educate consumers about the value of reusable packaging, highlight its advantages and provide rewards for choosing reusable options. Initiatives such as deposit-refund programmes, which reward customers for returned packaging, can encourage participation and modify consumer behaviour change.

Reusable packaging has become a powerful problem-solving method to deal with the environmental and economic drawbacks of using single-use packaging. It can lead your brand by significantly reducing waste, protecting natural resources, and lessening your carbon footprint. Furthermore, economic advantages like reduced expenses and improved supply chain efficiency provide an option for progressive business. A brand may improve its sustainable initiatives, further enhance packaging systems, and bolster its competitive advantage in a market that is becoming more environmentally concerned by adapting the latest innovations and trends in reusable packaging.

5.3.3 Recycle

Huge waste is produced by the apparel industry which can be recycled. This process will encourage a more circular economy. The type of packaging used for clothing items can differ in material and size. The materials used include paper, cards and poly bags. Translucent plastic materials in the form of polyvinyl chloride (PVC), and polyethylene terephthalate/ polyethylene (PET/PE) may also be used. Also, there is nonwoven low-density polyethylene (LDPE) perforated plastic packaging for apparel.

Recent developments have, however, made it possible to use eco-friendly and sustainable materials. These types of packaging materials help to reduce a brand's environmental footprint by removing polybags from its supply chain. For some brands, it is important to reduce, reuse, and recycle and it is the way to go. Not all products are of the same size so one can reduce their packaging and then also look at recycling them. To elevate the brand's ethics, one should reduce packaging size, select recyclable materials, and strive to reuse them. Initially, brands must be aware of what the consumers do with the packing materials. Whether they reuse and recycle it or simply throw it in the trash bin. "The sustainability of packing materials" has become one of the most significant terms used by social activists in recent decades. It helps the environment to reduce a fraction of the abundant garbage that is generated across the globe. Mostly all packing materials are recyclable. The various materials that are recycled to a larger extent are boxes with gummed paper tape, boxes with plastic packing tape, printed boxes, poly mailers, kraft paper mailers, padded mailers, bubble wraps, crinkle paper, tissue, and butcher paper. One should wisely choose the material packing, such that it should shift the environment in a positive direction. Choosing eco-friendly packing products is also one of the most advisable practises.

5.3.4 Refuse

There is a proverb that says 'Prevention is better than cure'. When we relate this, instead of using non-biodegradable and finding a cure for a sustainable world, we can start to refuse the harmful things that affect the environment. It is a good strategy for a brand to lead their company more sustainably. Here are a few things listed below that the company can choose to avoid:

Excessive Packaging: Many consumers are increasingly aware of the environmental impact of excessive packaging. Companies should refuse to use bulky or unnecessary packaging materials to minimise waste caused by those packaging materials and preach sustainability.

Single-Use Plastics: Avoid the use of single-use plastics materials which are used for packaging. Instead, switch to more eco-friendly options available to reduce plastic waste and contribute to environmental conservation efforts.

Non-recyclable Materials: Refuse to use materials that are difficult to recycle or reuse. Materials that cannot be recycled or reused can cause landfills. So, apparel companies and consumers must say no to materials that cannot be recycled to align with the goals of sustainability.

Complex Packaging: Simplifying packaging designs and reducing the number of components (like inserts or additional layers) can streamline production processes, lower costs, and potentially reduce the overall environmental footprint of the packaging.

Non-biodegradable Materials: Some brands may refuse to use materials that do not degrade naturally over time, such as certain types of foam or coatings, in favour of biodegradable or compostable alternatives.

These are some of the things that the apparel company should refuse to use to ensure sustainability and lead to a circular economy.

5.3.5 *Repurpose*

Repurpose in apparel packaging is the process of finding alternative uses or extending the lifecycle of the packaging materials which is used in the packing of apparel materials. Instead of discarding these waste materials, repurposing finds new ways of using them. This method helps in reducing wastage, protecting resources, and reducing environmental impact. By utilising seaweed, mushroom roots, and cornstarch, some of the brands provide sustainable packaging solutions that offer biodegradable features and a reduced ecological footprint without compromising on the quality and aesthetics that consumers expect from their packaging. Algae and Mushroom mycelium are two of the innovative materials being developed for eco-friendly packaging solutions. These materials contribute to a closed-loop, zero-waste ecology by providing the stiffness and protection required for transit and by naturally disintegrating upon disposal.

A significant step towards ensuring fashion packaging's afterlife is as considerate as its contents are the creation of compostable film and coatings. Under composting conditions, these films break down and become nutrient-rich soil that is good for plant growth (Ibrahim et al., 2022). These are some of the ways to repurpose that encourage individuals and apparel industries to rethink how they use and dispose of packaging materials, contributing to a more environmentally friendly approach to a sustainable environment.

5.3.6 *Rot*

Rot in apparel packaging is making use of things that are compostable which are bio-based. Through biodegradation, packing materials in landfills can be broken down into their constituent parts by nature. Conversely, compostable materials break down in conditions that are specially designed to control variables including temperature, humidity, oxygen, and bacteria. This will reduce landfills which cause environmental pollution as the biodegradable things are used. There are a lot of benefits for apparel companies and consumers for using biodegradable things in packaging. Some of the benefits are listed below.

Zero toxins and allergens: biodegradable packaging eliminates chemicals such as phthalates (commonly used in plastic packaging) which cause health issues like cancer and asthma. It is good for the environment and also does not contain toxins.

Reduce expenses: Mostly biodegradable packaging is recycled and reusable so investing in sustainable packaging will be pocket-friendly and free up resources that can be repurposed somewhere.

Efficient disposal: Sustainable packaging is labelled as biodegradable, compostable or recyclable which helps the consumers to dispose of the packaging wastes efficiently.

5.3.7 Rethink

Rethinking is providing new ideas and alternatives with new designs and features that can help to lead the goal of sustainability in packaging materials. There is already a lot of ideas that have been created like using peanuts (Styrofoam) for biodegradable packaging, corrugated bubble wrap, recycled cardboard, bio-plastics, inflatable air pillows using biodegradable materials, mushroom packaging, cornstarch packaging, seaweed packaging, and organic fabrics. Apparel brands should also encourage consumers to rethink once whether the product they get into the hand is fully eco-friendly.

5.4 Sustainability Initiatives

Many brands and retailers use the 7Rs to make the packaging materials sustainable. Some of the packaging solutions provided by industries are mentioned here.

5.4.1 Re-Pack

Re-pack focuses on providing sustainable and reusable packaging solutions. Online retailers and their clients can reuse and return their delivery packages of bags with the help of a re-pack.

This organisation has received lots of awards in design relating to sustainability. Services provided by re-pack are,

- Custom-branded packaging option
- Take-back systems
- Consultation for circular packaging solutions.

Companies like Levi's and H&M have partnered with re-pack.

5.4.2 *Noissue*

Josh Bowden, the owner of Noissue provides branded packaging for other brands and companies. Noissue offers many sustainable products like tissues, stickers, shipping boxes, takeout containers, cups, cards and tags, standup pouches, bags and totes, coffee bags, and many more. Their community is called eco-alliance. The "Noissue eco-packaging alliance", initiated by them presents a business that provides carbon-neutral packaging options to its members. It consists of business people, brands who choose to be sustainable and more. For every order placed, a tree is planted by this organisation in places endangered by deforestation. Members of this organisation contribute to this reforestation process with the issue organisation.

5.4.3 *Prana*

Prana is an American clothing company that started a movement called "The Responsible Packaging Movement" in 2020. It was started to reduce plastic pollution, mainly single-use plastics. More than 150 brands have joined this movement and the non-profit Canopy has also joined this movement. They have monthly webinars and round tablet conferences where they together explore ways to make more sustainable choices. These brands work together to eliminate single-use plastics and source only FSC-approved wood from properly managed forests and to preserve ancient forests.

5.4.4 *7Rs Implementation by Apparel Brands*

The various 7Rs initiatives taken by luxury and fast fashion brands are discussed here.

5.4.4.1 Gucci

Reduce Gucci was the first luxury brand that gave sustainability a priority and started working towards it. They wanted to reduce the sourcing of materials that were unsustainable and harmful to the environment. Gucci also released a new packaging line where the packaging is 100% recyclable. They also were adamant about reducing plastic usage so minimised using plastic materials in the packages which was followed by the use of water-based inks to reduce the pollution of water bodies with toxic inks. The brand has almost reduced all single-use plastic in online and offline packaging and aims to reduce it fully by 2030.

Refuse Gucci as a brand had to turn to paper as it reduced the usage of plastic materials. But even then Gucci refused the sourcing of wood and paper inappropriately. Gucci has partnered with Forest Stewardship Council (FSC) which is a global standard for forest sustainability. It protects the forests and allows sourcing of wood and paper only from properly managed forests.

Recycle We saw that the packaging was made of 100% recyclable materials, Gucci also upgraded the package handles to 100% recycled polyester from a conventional one. The handles are knotted together to prevent the usage of adhesive which contains toxic chemicals and causes a negative impact. Gucci dust bags and suit bags are made from 100% recycled cotton and 100% recycled polyester is used for the decorations. Hangers in the retail stores are made of recycled polystyrene.

5.4.4.2 Chanel

Reduce Chanel uses glass jars to package their makeup products. The weight of these glass jars underwent a significant reduction in their weight to reduce the environmental impact created by them. Their glass bottles' weight has been reduced by about 30% in weight. A particular jar "crème revitalisante jar" has been made to be nearly half as light as its previous weight. An advantage of reducing the weight of any product or material is that when disposing the environmental impact will be less as it takes only half the time to decompose than the heavy-weight materials.

Rethink Chanel has collaborated with Constellium and a luxury packaging specialist g.pivaudran to launch a new solution specially for its mascara "Le Volume". About 20% of post-consumer recycled aluminium is used in the package for decoration purposes. In Chanel's "Chance eau de toilette", an aluminium strip was used in the decoration of the bottle instead of previously used metal brass. As aluminium can be recycled as many times as possible compared to the brass metal strip, this has brought a reduction in waste material disposal.

Repurpose Chanel adopted camellia seed shells in the lids of their containers. Camellia seed shell is a next-generation alternative to conventional packaging material. Next-gen alternatives include materials that are abundantly present in the world but are generally considered waste and thrown away or burnt. When these materials are repurposed to replace plastic or paper, the waste generation is widely lowered and thus automatically the environmental impact is reduced.

Refuse Chanel refuses the use of heavy-weight packaging materials and always prefers materials that are lighter in weight. Like Gucci, Chanel refuses to source unsustainable materials. It uses wood that is FSCTM5 certified and uses only water-based ink for the decorations.

5.4.4.3 H&M

Reduce A few goals that H&M has been working on are,

- By 2025,
- H&M aims to reduce plastic packaging materials in their brand by at least 25% and the materials used at that time are aimed to be of the reusable sort or if disposed of, they need to be 100% recyclable.
- By 2030,
- 100% packaging has to be recyclable and made from materials that are post-consumer recycled materials.

Rethink H&M is supporting the Canopy pack4good initiative. Canopy pack4good initiative expects the company to, source FSC-certified wood, only use recycled content to make packages, use post-consumer waste materials and recycle them, practising new packaging strategies such that only less content is used.

Reuse H&M makes sure that their online paper packaging is reusable. After using it for a long time if the package gets torn or distressed, it can be easily recycled as it's made of 100% recyclable materials which can be used again and again.

5.4.4.4 ZARA

Rethink Zara initiated a movement "green to pack", which is a movement that expects the brand to use sustainably sourced materials and recycle their packaging materials so that they can be used repeatedly.

By 2016 Zara achieved,

- 10% thinner plastic bags.
- Reusing packaging boxes and millions of hangers.
- Using recycled cardboard boxes from their brand and others.

Zara has also recently joined the "canopy pack4good" initiative to protect the forests and properly source for paper.

Reduce The "Green to Pack" initiative has given very good results to the brand. This initiative has helped the brand to reduce around 80% of the paper used in packaging and millions of dollars which was saved can be used elsewhere. The "BRINGYOURBAG" initiative introduced by the brand encourages consumers to bring their bags while shopping. This is implemented by charging the consumers for the paper bags given in the retail stores, where prices increase according to the size of the bag. Around 47% of paper bag consumption has been successfully reduced.

Repurpose The recyclable paper content is further replaced by next-generation materials such as cereal straws, hemp stalks, jute, tomato stems, etc. These can be considered as replacements for silk, wool and Materials like that as these are environmentally better than those.

5.4.4.5 Aditya Birla Groups

Rethink Aditya Birla has collaborated with a circular apparel innovation factory (CAIF) and produced biodegradable polybags. Through all these methods, this group has achieved 86% sustainable packaging.

Repurpose Here, around 3,39,675 tons of non-recyclable plastic waste is used as alternative fuel and 1,24,070 tons of municipal solid wastes are used as an alternate fuel. Its goal is to use 100% sustainable packaging materials by 2025.

Recycle Aditya Birla groups recycle all their packaging wastes such that 0% of waste goes to landfills. Aditya Birla Group has attained a status where it contributes no waste to the landfills across all of its facilities which is a significant achievement by the company.

5.4.4.6 Prada

Rethink In 2023, the group was able to complete a transition from polypropylene to microwave cardboard for sneakers boxes and plastic is replaced with paper and knitted bags. This reduced the plastic consumption of the company to about 10%.

Recycle This group only purchases wood and paper that are FSC-certified. They try to maintain the properly sourced wood to always be above 95%. The fashion pact has a B2C has a B2C target for 2025, which states that more than 56% of the packaging of B2C must be made of 100% recycled plastic. Prada achieved that target 2 years earlier.

Rethink Prada has introduced a re-nylon project where a new line of regenerated nylon known as ECONYL is released. The company has partnered with Aquafil and Aquafil manufactures ECONYL from plastic waste from oceans and textile wastes. Prada made all its plastic products to be replaced with ECONYL in 2021.

5.4.4.7 Shein

Recycle Shein has taken it into account to use recycled cardboard boxes and biodegradable mailers instead of plastics. It also emphasises using water-based inks for printing designs instead of chemical ones for packaging covers to reduce water pollution.

In 2022, Shein began using recycled PE plastic express bags for the parcels that have to be sent to the UK. These express bags contain 50% PE plastic. PE plastics can be recycled continuously so the raw materials needed are comparatively much less than for virgin materials.

Reduce By using PE plastic, the raw materials used are reduced, the water wastage is reduced to a distinct amount, and the plastic waste is also seemingly reduced.

By the end of 2024, it calculated that 319 tons of virgin plastic is saved only through Shein packaging. Three tonnes of virgin plastic is saved monthly through the PE plastic initiative.

5.4.4.8 Myntra

Reduce Single-use plastic has been eliminated 100% in Myntra's packaging and is replaced with green packaging. All the bubble wraps, plastic cello tapes, and poly covers are replaced with sustainable materials such as paper covers and boxes, etc. Myntra has completely discontinued the use of plastic polythene covers or plastic outer packaging. This has reduced over 1.4 tonnes of plastic in 2021 and about 115 tons of plastic in 2024, according to Myntra.

Rethink Myntra has partnered with Canopy, an NGO so that it can be made sure that the used packaging is collected from properly maintained forests and not through deforestation.

It has also partnered with "Better Cotton Initiative" which is a programme that takes the initiative to make global cotton production sustainable.

5.5 Tools Used to Measure Sustainability in Packaging

There are various tools or software used to study the efficacy of the sustainable methods of packaging used by industries. Those tools are discussed below:

5.5.1 PIQET

Packaging Impact Quick Evaluation Tool

It is an online business tool that is specially designed and used for the Packaging Industry. It is a Streamlined Software tool that is used to calculate the environmental impact and resource consumption of different packaging materials used in the different packaging options. It uses software which is ISO 14040. The driving factor for PIQET is the approach to Life Cycle Assessment (LCA). The datasets used in this are developed in Simapro Life Cycle Assessment (LCA) which is based on the life cycle inventory (LCI) which consists of 18,000 unique datasets. The databases are updated regularly. It has a transparent database. It has a large database network for which data are taken from background data. It has a wide variety of sustainable impact products which helps us to evaluate the packaging materials to give us the alternate product which is Sustainable and environmentally friendly. The Estee Lauder Companies Inc. is a personal care product brand which is based in New York who is working with the PIQET to get help in the end-to-end product lifecycle like from ingredient selection and sourcing to product formulation and packaging. Estee has elevated to new sample packets which are made using sustainable materials which has increased their brand value. Currently, PIQET has developed two new sustainable materials for packaging which are Mycelium which is developed from fungus which is a bio-based polymer and Adhesive oil based which is currently being added to their new data for LCA calculation.

5.5.2 Pickler

It is an environmental footprint calculation software for packaging products. It was founded by Koen de Beer and Daan Van Han who had a background in innovation and venture building. Before starting this company, they both worked with many packaging companies after finding out that the whole packaging process was not sustainable, they found this company to develop sustainable packaging. They have joined hands with Sustainable Impact Merits (a technical university) to develop an instant scalable and transparent Life Cycle Assessment (LCA) for packaging which is known as the 'Fast Track LCA'. They have an impact database which is known as Idemat which is constantly updated for the usage of customers. This helps to empower packaging resellers to calculate and reduce the impact of their packaging material and promote the Anti-Green Washing guidelines. Purepack which is a sustainable packaging company uses Pickler to provide calculated environmental footprint of packaging proofs to their customers which has not only improved their brand name but has also increased the trust of their consumers.

5.5.3 *Fast Track Life Cycle Analysis (FT LCA)*

It is a tool that helps companies and designers assess the environmental impact of a product or service so that they can produce an alternate product or service that does not damage the environment.

5.5.4 *Trayak*

It was found by Prashant Jagtap. It is a cloud-based platform that helps its users to assist and analyse the sustainability of packaging products. It offers a sustainable platform and life cycle assessment (LCA) that allows its users to design their product and view their product's lifecycle. It helps companies to create an entire portfolio using sustainable strategies. They intend to show that using sustainable strategies in current manufacturing and designing is easy.

5.6 Conclusion

The apparel packaging industry has a significant opportunity to reduce the environmental footprint caused by them by using sustainable packaging. They can switch to eco-friendly materials which help in minimising the packaging waste and designing a circular system. Companies can reduce their carbon dioxide emission and decrease pollution by using the 7Rs and sustainable packaging tools. Using sustainable packaging materials will help in enhancing the brand's reputation, customer loyalty, and cost saving. As consumers are beginning to get environmentally conscious, it is no longer a choice to be sustainable but a need for a long run in the business. In this chapter, the various functions of packaging, materials used in packaging, forms of apparel packaging, concept of 7R, its usage in apparel brands for packaging materials sustainability, and tools or software used for evaluating the efficacy of the use of sustainable packaging materials are discussed paving way for a better understanding of how retailers can turn to sustainable practises.

References

Chamas, A., Moon, H., Zheng, J., Qiu, Y., Tabassum, T., Jang, J. H., Abu-Omar, M., Scott, S. L., & Suh, S. (2020). Degradation rates of plastics in the environment. *ACS Sustainable Chemistry & Engineering, 8*(9), 3494–3511.

Dhiman, V., Kumar, A., Chaudhary, A., Singh, J., Goyal, A., & Kanoungo, A. (2022). Utilization of plastic waste and its management: A mini-review. *ECS Transactions, 107*(1), 9127.

Gray, M. (1939). The history and development of packaging. *Journal of the Royal Society of Arts, 87*(4511), 633–658.

Ibrahim, I. D., Hamam, Y., Sadiku, E. R., Ndambuki, J. M., Kupolati, W. K., Jamiru, T., Eze, A. A., & Snyman, J. (2022). Need for sustainable packaging: An overview. *Polymers, 14*(20), 4430.

Jestratijevic, I., Maystorovich, I., & Vrabič-Brodnjak, U. (2022). The 7 Rs sustainable packaging framework: Systematic review of sustainable packaging solutions in the apparel and footwear industry. *Sustainable Production and Consumption, 30*, 331–340.

Pollman, E. (2019). Corporate social responsibility, ESG, and compliance. *Forthcoming, Cambridge handbook of compliance (D. Daniel Sokol & Benjamin van Rooij eds.), Loyola Law School, Los Angeles Legal Studies Research Paper*, (2019-35).

Rhodes, C. J. (2019). Solving the plastic problem: From cradle to grave, to reincarnation. *Science Progress, 102*(3), 218–248.

Royal (2024). *How much waste does the fashion industry produce?* [Online]. Available from https://royalwaste.com/how-much-waste-does-the-fashion-industry-produce/#:~:text=As%20of%202023%2C%20the%20fashion,contributions%2C%20which%20exacerbates%20environmental%20issues. Accessed on 20 June 2024.

Chapter 6
Analysing Different Strategies for Adopting Sustainable Apparel Retail Practices

R. Surjit, H. Shanruthi, V. Sruthi, and K. S. Tarunvalavan

6.1 Introduction

The concept of purchase majorly involves the customer and retailer interaction that eventually ends up profiting the retailer, and the whole supply chain that is invisibly existing behind the retailer. Making a person buy a product is so much more common. But understanding the market and making them buy what they want is what is called real marketing. However, over the years the evolution of mankind and their lifestyle is tremendous. Nowadays people not only buy for their needs but also buy for their desires. With our planet increasing its temperature, and being exploited each day, a significant growth of awareness has been instilled in the minds of people. This pushes them to become more socially responsible. It has impacted all aspects of their life which majorly includes their purchase, consumption, and selling. As an enormous preference for eco-friendly and sustainable products grew among people, retailers have accordingly made it to a point to sell them the same. However, it is not an easy task to do. For any product to be sustainable, the concern should begin right from the person who produces the raw materials, which is not so easy. Especially, with major concern in the apparel sector, customer's consciousness towards buying a sustainable garment is growing to become inevitable shortly. Moreover, for the fact that there are people who are still unaware of the importance of sustainable apparel retailing, this chapter provides eye-opening insights that highlight the same. On the other hand, for retailers, it is still a huge task. All the above-mentioned factors can be addressed and accomplished through developing various strategies to bring them into action and adopting them properly such that

R. Surjit (✉) · H. Shanruthi · V. Sruthi · K. S. Tarunvalavan
Department of Fashion Technology, PSG College of Technology,
Coimbatore, Tamil Nadu, India
e-mail: rst.fashion@psgtech.ac.in

S. S. Muthu (ed.), *Sustainable Apparel Retailing*, Sustainable Textiles: Production, Processing, Manufacturing & Chemistry,
https://doi.org/10.1007/978-3-031-83875-0_6

they end up being beneficial. This chapter majorly involves the overall analysis of these strategies from a much broader perspective. Furthermore, having an eye on the sustainability of the outcomes, these strategies generate is discussed as well. Apart from these strategic measures that retailers take to attract the interest of consumers, they have various other responsibilities that are highly associated with their relationship with suppliers, which is all a part of this chapter. Knowing how these strategies, when implemented, reflected in better sales is equally important. This chapter covers all these areas along with providing different ideas that give scope for future analysis on apparel retailing.

6.2 Importance of Sustainable Apparel Retailing

Among the numerous different industries that exist, the concept of sustainability has developed from a small fringe to very great mainstream issues to not only the manufacturers, but the marketers, managers, and even stakeholders for that matter, for the past few years. The concern towards the textile and apparel industry in aspects of societal and environmental well-being is often aloof, even after endless steps taken to create worldwide awareness. But, the reality is terrifying and is posing various serious threats to nature.

According to research, the textile industries all over the world are recorded to be completely responsible for the emission of 2.1 billion tons of greenhouse gases in the year 2018. The major problem at the customer end is that they do not seem to effectively translate their green attitudes into subsequently beneficial green actions. Studies also say that customers who fall under the age category of above 35 are not so supportive of sustainable fashion, they just tend to be neutral in their opinions about it (Rausch et al., 2021). A full-fledged sustainable fashion retailing can result only if both the retailer and the customer equally understand that sustainable apparel is the need of the hour in the prevailing environmental conditions. For the retailer, it brings with it the entire set of difficulties that is faced by the whole supply chain. It is the retailer who coordinates with everyone and brings together different products to provide refined product options so that customers can choose what they want from those options. Manufacturing a sustainable product is not so common as it requires proper concern and follow-ups right from bringing in the raw materials that are eco-friendly and sustainable as well.

In the entire supply chain, it is the retailer who serves a proper connecting system between the suppliers who form the upstream and the consumers who constitute the downstream. Thus, they serve as a key in driving the circular economy. As part of this, the products at the consumers' side are obtained at the end of their life stage and again given to the supply chain as a raw material, to have the continual flow of reuse going. By significantly co-relating the consumers and suppliers, the crucial retail initiatives have proven to be useful in the reuse, reduce, and recycle concept of sustainability (Vadakkepatt et al., 2021). The retailers play a critical role in the supply chain which makes them act as "ecological gatekeepers". The activities of these retailers can highly influence environmental and social sustainability in various aspects. Insufficient research about such an important key role player in the supply chain, gives more scope for further analysis and data procurement, paving ways to incorporate more detail in this chapter. Moreover, the fashion industry has been continually attracting the complete attention of various researchers over for so many years. The ever-growing complexity in this dynamic context has to a large extent increased their interest in this area.

Fashion constitutes one of those industries in which the market constantly undergoes rapid changes, which requires more responsive and flexible players. However, as retailers stand in the frontier, their flexibility and responsiveness in the supply chain is highly crucial. Retailers are highly expected to minimise the waste generation and emissions of effluents, through continuous improvement in all their internal operations. However, sustainable retailing includes two main practices, namely green store operations and green transportation. Green store operations refer to all those retailing operations that facilitate energy conservation, recycling, or waste reduction. The green transportation refers to the eco-friendly green products. It is also said that sustainable retailing includes facilities management, green product sales, customer-facing initiatives, and the retail supply chain (Yang et al., 2017). It is to be noted that the initiative of sustainable retailing not only benefits the environment and sustainable well-being but also benefits the business itself to a large extent. Sustainable retailing in general, is all about the selling of various products and services in such a way that it reduces all the possible negative social and environmental impacts while at the same time working towards maximising the positive aspects. It comprises a variety of elements right from sourcing materials that are sustainable to properly supporting fair trade practices. They strive to their fullest to operate such that they claim to be socially and environmentally responsible. In most cases, sustainable retailing is carried out to ensure that a product or service offered by a company is environmentally friendly and that they don't pose prominent negative impacts on the planet. Over the years, with increasing environmental concerns and to highlight the good ways in which the corporation gives back to nature that has been exploited to create itself, it has taken up different Corporate Social Responsibilities (CSR) initiatives. With such increased awareness of devastating climatic effects, a change grows in the mindset of people who seek to reduce their part of the impact they create on the environment. This has led to a state where almost 70% of Gen Z and millennials are ready to pay more and afford costly products if they are guaranteed proper sustainability.

Nearly, half the global consumers are willing to buy a product or obtain a service from a brand or retail outlet that has a clear commitment towards achieving sustainability. As per numerous surveys, research, and analysis that was conducted based on such customer behaviour, it is predicted that it is a sure event that this preference of customers over sustainable goods and entities will only continue to grow over the years, making it equally sure that sustainable retailing will soon become a prerequisite rather than something that the brand or store can boast about to gain more popularity. So, it is high time that everyone realises that sustainability is the future of not only business but all sectors for that matter. Moreover, the need for implementing sustainability is prominent for a fact that consumers are constantly in search of brands or stores that tend to reflect their values. Also, it is likely to win their loyalty for the brands or organisations that invest more in practices that are sustainable and are working towards the well-being of the global economy, society, and environment. The advancements and more prominent access to massive information are impacting people, which make their concern about reducing environmental exploitation grow day by day. Now the situation is such that, customers are not only conscious about buying sustainable products, but are also equally concerned about sustainability in its packaging and delivery. The recent years have recorded a weighty shift in customer loyalty towards more sustainable brands and entities that are said to be carrying out ethical business (Diversitech, 2022). In such a complex world full of global supply chains and manufacturing, the customers to a larger extent seek the help of the retailers to adopt a more sustainable lifestyle. According to various research studies, many brands and retailers are working on developing themselves as per customers' preferences. This includes data that says 64% of customers prefer reduced packaging, 50% of them prefer to be informed about details on how to recycle the product, and 46% of them are said to seek certain crucial clarifications regarding how the products are being sourced. Thus, the proper importance of incorporating sustainable practices into business, especially retailing is highlighted.

6.3 Green Retail Practices

The studies on the main fashion business brands that are associated with green retail practices and sustainable retailing, the prevailing sustainable business practices in the fashion retailing industry, and the trends and prospects of sustainable retailing in light of the quick expansion of e-commerce in the fashion sector are discussed here. In the textile and apparel sector, "green retail practices"refer to a range of ethical strategies intended to reduce harmful environmental effects and promote sustainability in all three terms, i.e., environmental, social, and economic (Yang et al., 2017). There are many key practices followed currently to achieve sustainability in retailing. They are:

- Utilisation of sustainable materials
- Cradle to cradle fashion
- Sustainable production of materials
- Operations for retailing sustainably
- Transparency and certifications
- Consumer education and engagement
- Waste management approaches

6.3.1 Utilisation of Sustainable Materials

Reducing the environmental impact of the fashion industry is largely dependent on the use of sustainable materials. For instance, organic cotton is produced without the use of artificial fertilisers or pesticides, which lowers pollution of the land and water and increases biodiversity. As the world's second-largest pollutant in the textile and apparel industry, the utilisation of sustainable materials in this area has received a great amount of importance. The process of the conversion from raw materials to valued added garments has added significant environmental, social, and economic implications on the planet and this has happened especially in low-cost labour countries like India, Bangladesh, and China (Shen et al., 2017). Similarly, compared to conventional cotton, natural fibres like hemp and bamboo aren't the only renewable. Those fibres require very less water and chemical resources to grow. The use of these sustainable fibres can promote healthier ecosystems and lower the ecological impact of apparel production. Furthermore, manufacturers are increasingly using advanced materials like Tencel, which is made from wood pulp that is sourced ethically, eco-friendly, and produced using a waste management system that eliminates waste and emissions.

There are many sustainable raw materials available to cater the apparel and textile production including organic fibres and recycled materials (Muthu, 2014). Sustainable fashion also includes a lot of recycled materials. Many regenerated fibres are now made from textile wastes and also exhibit an excellent property of longevity in terms of quality and usage. This promotes the development of a circular fashion economy. These materials contribute to reducing the growing issue of textile waste while also reducing the need for fresh resources. The apparel retail industry can move towards a more ethical and eco-friendly manufacturing method by inculcating recycled and sustainable materials, which will always encourage more ethical and sustainable approaches to apparel retail and consumption.

6.3.2 Cradle-to-Cradle Fashion

The cradle to cradle is a lifecycle model in LCA. LCA refers to lifecycle assessment. It is a method or a process that is employed to measure the environmental impact of a product from its production to the end. LCA studies the environmental impact of a product from cradle to grave (till the waste management). To reduce waste and increase the lifespan of items, circular fashion in retailing aims to establish a closed-loop system where clothes are designed, made, consumed, and then recycled or repurposed. This is a cradle-to-cradle approach where the products are recycled, reused, and repurposed to the maximum possible extent and not thrown as a waste. The retailers who support circular fashion encourage the use of easily recyclable or biodegradable sustainable materials. The promotion of vintage and used apparel also contributes to circular fashion in retailing by extending the lifespan of items and lowering the need for new manufacturing. For prolonged existence, the apparel products must be designed with maximum resource effectiveness, non-toxicity, biodegradability and reusability, and ethics in the mind of the stakeholders (Rathinamoorthy, 2019). In addition to reducing waste, circular fashion in retailing also encourages customers to embrace more sustainable consumption practices, which in turn supports a more robust and sustainable fashion business.

6.3.3 Sustainable Production of Materials

In the retail industry, sustainable production refers to the use of methods that reduce the negative effects on the environment and encourage resource efficiency through the supply chain. The first step in achieving this is implementing energy-efficient technologies in industrial processes, such as lowering dependency on fossil fuels by utilising renewable energy sources like solar and wind power. Furthermore, water conservation strategies, such as low-water dyeing procedures and closed-loop water systems, greatly lower water usage and pollution. Retailers may lessen their environmental impact even more by emphasising the use of non-toxic products and making sure they are disposed of safely. These procedures not only improve operational effectiveness and save long-term expenses, but they also support environmental sustainability.

6.3.4 Operations for Retailing Sustainably

Sustainable operations in apparel retailing focus on minimising the environmental impact of store operations and logistics. One key aspect is the design and construction of eco-friendly retail spaces. This includes using sustainable building materials, such as reclaimed wood and recycled steel, and incorporating energy-efficient lighting to reduce energy consumption. There are environmental impacts such as natural resource extraction, water, and energy used in production, pollution, transportation, use of the product, and finally disposal. This follows all along a product life cycle (Medium, 2018). Another critical component is 'green logistics'. It gives rise to environmentally friendly efficient transportation for a sustainable future when green (code for the environment) and logistics are put together (Rodrigue et al., 2017). This involves optimising transportation to lower carbon emissions, such as using electric delivery vehicles and selecting shipping routes that reduce fuel consumption. By implementing these sustainable practices into their operations, apparel retailers can certainly eliminate their ecological footprint and promote a more sustainable apparel retail sector.

6.3.5 Transparency and Certifications

Transparency in business is the act of providing complete details about goods and services, as well as the goals and beliefs of the organisation. The goal of product traceability is to show the origins, contents, and manufacturing processes of items. Transparent businesses provide their consumer base with simple access to information about everything from raw materials to sourcing to the point of sale.

6.3.6 Consumer Education and Engagement

Retail consumer engagement refers to how a brand or a retail store engages with its consumers across touchpoints and channels, both brick-and-mortar and digitally. It influences the loyalty of the relationship a consumer develops with a.

6.3.7 Waste Management Approaches

Waste management in the apparel retail sector involves a holistic approach aimed at lowering waste and achieving sustainability throughout the product lifecycle. Waste management approaches such as the 3Rs (Reduce, Reuse, and Recycle) and lean manufacturing techniques (5S) could be readily used to reduce waste inside the retail store. It finally encourages a circular economy in fashion (Islam, 2021).

6.4 Strategies of Sustainable Apparel Retailing

The competition among the retailers in the apparel industry remains unimagined nowadays. In that scenario, to sustain their business in the market, retailers have to attract their customers uniquely (Aydin et al., 2023). There comes the growing sustainability concern in play. For retail to be sustainable, it is obvious that its whole supply chain of the product should be sustainable. To make that happen, there are various strategies that the retail stores are following to satisfy their growing sustainable-based customers. Also, there are studies and analysis that have been done on the existing sustainable strategies. One such major strategy is explained below:

6.4.1 QFD Method

Quality function development is the method of transforming the customer's demand or needs into quality-designed products, especially in manufacturing. This method is ultimately used to increase consumer satisfaction ensuring to meet their expectation. It makes the customer demand into the process or services mainly in terms of quality. Rather than producing in tons and keeping it for sale, it is better to produce in the consumer requirement. There are certain steps in how we can apply the QFD method into play,

(i) Identifying the customer requirements: The first step is to know about the consumer needs and requirements, i.e., the voice of the customer. This can be identified by conducting surveys, interviews, focus group discussions, questionnaires, getting feedback, analysing the feedback, and online reviews.
(ii) Customer requirements to Design requirements: By using the collected data from customers, their requirements are made to check with the product attributes to enhance the product. The attributes can be of anything like colour, fabric quality, fit, etc., and also determine the company's engineering characteristics to produce the customer needs as the Voice of the company.
(iii) Developing HoQ: HoQ is the House of Quality matrix where the customer requirements are listed on one axis and the technical requirements of the company on the other axis. This gives us a clear-cut view of the customer's needs and also helps to assign the rating for their needs based on our engineering capability.
(iv) Benchmarking and Competitive Analysis: After that, the retailer can compare the product specifications with the competitors' products in all terms to identify the product's strengths and weaknesses and to produce the same effectively and efficiently so that the product's market can increase.
(v) Set Priorities and Allocate Resources: Based on the above-said steps, now the retailer has an idea of the product that is to be prioritised first with the existing capabilities and customer wants, accordingly allocate the resources to produce the product.
(vi) Develop and Test Prototypes: For the prioritised model, a sample should be developed as per the customer requirements, and trials can be conducted with the customers to ensure that the prototypes have met their needs or not.
(vii) Refine and Finalise the design: On getting the feedback from the prototype, re-design or re-engineer the product and finalise the design specification for mass production.
(viii) Implement Production and Quality Control: On the implementation of the bulk production, adhere to the product's required quality standards and parameters that are mainly to meet the customer demands. Main quality consistency in every product that is produced.
(ix) Continuous improvement: Even after production and starting sales, continuously gather customer feedback and monitor market trends. Try to make ongoing improvements to product designs and processes.

6.4.2 Benefits of QFD in Apparel Retailing

- Enhanced Customer Satisfaction: By closely aligning product features with customer desires, QFD helps in creating products that customers love and prefer.
- Improved Product Quality: The systematic approach ensures that all critical quality aspects are considered and addressed during the design and production phases.
- Reduced Time to Market: Clear specifications and priorities help streamline the development process, reducing the time needed to bring new products to market.
- Cost Efficiency: By focusing on the most critical customer requirements, resources are used more efficiently, potentially reducing costs associated with redesigns and returns.
- Competitive Advantage: Products that better meet customer needs can differentiate a brand in a competitive market, leading to increased market share.

Fulton and Lee (2012) conducted a study on sustainable apparel online retailers to identify their characteristics and evaluate sustainable initiatives in the complete supply chain process. They have started to gather data and facts by checking the company's profile in the 'About Us' to know their policies and checking on the products that they offer. In that, mostly all the companies have provided only their history. To attract sustainably concerned people, websites started to put environmental certifications like carbon trust to ISO certifications. Some companies have provided the method of production like organically made, home-made, handcrafted, mass-produced, etc., on coming to their supply chain activities, the apparel online retailers started to shout out their sustainable organic raw materials in their websites made up of organic fibres, pesticide-free production of fibres and fabrics, recycled raw materials, biodegradable materials, and recyclable materials. In the case of the manufacturing aspect, the sustainable terms used were environmentally friendly dyes and eco-friendly screen printing, and also their usage of water for the processing had been mentioned on some websites. The distribution and logistics part is the major issue to be considered where the emission of GHG is comparatively high. So, the retailers started providing the details about the eco-friendly containers that they have been using for local and global shipments. And few companies have also tried alternate fuel biodiesel and programmes like bike borrowing. 4% of the retail shops have shared the details of their store's building materials which show the commitment to sustainability. A multi-channel retailer named Embodies uses eco-friendly materials such as cork floors, mannequins made from reclaimed Nike shoes, VOC-free paint, reclaimed lighting with CFL bulbs, fixtures made from FSC-certified wood, and modular wall units. Additionally, 4% of other websites address packaging details. Emommie is one such website, specialising in second-hand women's maternity clothing and serving as a platform for buying pre-loved garments. The retail stores of Patagonia accept old plastic bottles and polyester garments that are converted into recycled polyester fleece garments. The laundry and care were also mentioned on some of the websites like the use of organic detergents that will not allow the garments to degrade and wear out easily.

Marks and Spencer, a famous clothing retail store had made a suit that is claimed to be the most sustainable suit, which was made from Australian organic wool and recycled polyester. It is said that even the buttons and the zippers are recycled and reused. The main strategy that they follow is reducing packaging waste. Through their sustainable efforts, they have achieved a 20% reduction in the average packing weight. In the same way, retailers are ensuring the use of cargo ships and rails for bulk transportation purposes where it is more sustainable as compared to trucks and aeroplanes. Even though it takes time, it does not emit poisonous gases as compared to trucks and aeroplanes. Walmart in its 2012 Global Responsible report has claimed to achieve an 80% reduction in its overall waste.

Nowadays, a growing number of companies provide more repair facilities for their garments and footwear. In addition to that, retail stores also provide rental and resale services. Several popular fashion brands, such as H&M, Zara, Mango, and Marks & Spencer, have recently launched their recycling schemes. Alfonso Dominguez has introduced a rental service in his primary market, Spain. Meanwhile, Cencosud's Paris. Cl, which acts as an aggregator for brands like Vestua and Nostalgic, offers buying, selling, renting, and repairing of used items in its Paris department store. Additionally, mass retailer Primark has adopted a concession model with the launch of "Worn Well" vintage clothing corners in selected UK stores. If the pilot phase proves successful, this scheme could be expanded across Primark's network (LeRolland, 2023).

Dabija and Băbuț (2019) conducted research based on the Generation theory to analyse the behaviour of two consumer generations, i.e., Generation Xers and Generation Millennials (Gen Y). On various hypotheses, they have measured the behaviour of both generations and found that the apparel store's patronage behaviour of GenZ is influenced by the advertising, personnel, and assortment. They mostly like the explanations that are being given by the brands like why should we choose the particular brand, the specifications in it, and at the same time not leaving the sustainability behind. Sustainability is the major concern here in all areas and every aspect. Rather than producing only organic raw materials to attain sustainability, it's every person's responsibility to take care of the production, processes, and retailing to be sustainable.

6.5 Relationship Between Retailer and Supplier

In a supply chain, the suppliers and retailers are the ones who play major roles. The suppliers are the ones who supply to be precise. They comprise a business or a particular individual who is likely to provide goods and services to another company or an individual. While retailer is a person who does retailing business that sells a variety of products to the customers. The supplier networks are a structured framework that comprises the manufacturers and distributors from whom all the retailers purchase products so that they can sell products to the customers directly (Gillivan, 2024).

The co-existence of retailers and suppliers has brought about the idea of increasing collaboration between the two, which is commonly agreed upon by people in all good aspects. It is also said that proper collaborative efforts can significantly minimise the environmental impact, reduce the inventory throughout the supply chain, and hence eliminate the total end-to-end costs by up to 3%. It is also believed by 80% of the people that the collaborations between them are constantly growing at a larger pace, especially in the past 3 years. Despite this fact of growth, it is not being practiced so widely by everyone. Moreover, the entire achieved amount of benefits is usually incremental rather than transformational. As a consequent result, there are a lot of critical issues that pose difficulties not only to the retailer and manufacturer but the whole supply chain.

Some of these include the prevailing backlogs of fast-moving stock getting dumped and stored in the retail warehouses, false inventory holdings becoming a part of the same framework of the supply chain, minimal availability of products, especially promotional items, on the shelves and the truck fleets that are being underused. As per an analysis conducted by the Boston Consulting Group, it is said that the mobility of products, bringing them from factories to the store shelves is quite expensive part of the entire value chain. It also adds that it contributes nearly 9 to 17% of end-to-end costs. However, collaboration efforts aim to preliminarily streamline the inventory in a better way through proper forecasting, quicker order fulfilment, and practicing backhauling, involving picking up the goods from the supplier during their return trips once after the store-wise deliveries are done. Also, efficient utilisation of consolidation units and joint warehouses, ultimately reducing the overall costs. There are various challenges that these collaborations are prone to, which include those which are internal and external as well. There exist different cross-purpose operations that take place between the supply chain and merchandising functions, in any organisation that involves the retail and supplier for that matter. With the lack of proper communication and misaligned goals, they face numerous differences of opinion. It is quite common that without a proper system for information sharing, initiatives the collaborations are unlikely to succeed. This is majorly due to the continually changing short-term priorities of both the suppliers and retailers and it takes years to build a solid relationship.

One major thing that further complicates matters is the mistrust that has inherently been existing among the retailers and suppliers, which is the exact factor that keeps them from further working together. In most situations, there has been an imbalance of power that allowed retailers to offload the inventories and the logistic costs to the suppliers. This was a situation where shorter lead times for orders were desired along with much more frequent deliveries. Consequently, the suppliers end up in their attempt to recoup all those additional costs associated with it during their price negotiations with the retailers. Moreover, there were also ways in which retailers and suppliers tended to compete directly, mostly through private label-associated issues. Furthermore, suppliers, going a step forward, started finding other different platforms through which they can get theirs products sold directly to the customers, surpassing the conventional sales mechanism through the retailers.

Despite all the above-stated challenges and hindrances, a successful collaborative relationship between the suppliers and retailers can be brought about with significant investment in time and various other resources. This is exactly why, a proper commitment of senior executives from both sides is highly recommended and is crucial. Ensuring a systematic strategic alignment and defining various boundaries in their partnership, gives an escalation route to resolve all possible differences among them. This will in turn help maintain a perfect momentum even through certain highly critical hard phases that are inevitable. Apart from all these, one must choose the right partner to collaborate with, be it a retailer or a supplier. Also, making a proper long-term commitment can prove to be highly beneficial. Building cross-functional teams and having proper agreements on how the benefits are to be shared, are a few other factors that can help retailers and suppliers stay on good terms in the long run (Mercier & Battle, 2012).

6.6 Consumer Behaviour Towards Retail Transition

A very strategic and vital role is played by retail stores in distributing clothing goods to consumers (Timotius & Octavius, 2021). Purchasers today are increasingly knowledgeable and ethical, and they frequently look for brands that look upon their perception of ethical and sustainable production practices. Retailers are forced to change their practices according to the trends to match the evolving demands of their customers. In response, brands are being more open about their sourcing and production practices, securing certifications for sustainability, launching eco-friendly projects, and trying to catch up with the concept of a transparent supply chain. The retailers also interact with customers by educating them about the advantages of sustainable fashion and motivating them to make ethical decisions. Consumer engagement and interaction are done right the first time with good education policies and ethical behaviour of retail shops. In the clothing sector, consumer education is essential to promote and develop sustainable fashion and ethical purchasing.

The arrival of digital transformation has proven to be an inevitable force in the rapidly changing retail situation, changing established production processes, reframing customer expectations, and thus changing the dynamics between sellers and consumers (Sagar, 2024). Online platforms and social media play a significant role in this retail transition, as they provide an arena for clothing brands from luxury to ordinary goods, to communicate their stakeholder's efforts, and for consumers to share their valuable feedback about their experiences on using sustainable goods. Due to the growing use of smartphones, better internet connectivity, and convenience-driven purchasing behaviours, consumers have moved towards doing their shopping online. Lockdowns and other social distancing effects brought on by the COVID-19 epidemic have further grown this tendency by encouraging more people to use online resources (Linkedin, 2023).

Consumers are willing to pay a higher price to purchase sustainable apparel, prioritising quality and longevity over fast fashion. They also show greater interest in second-hand, vintage, and up-cycled clothing, shifting towards circular fashion. Since customers with greater incomes may easily afford sustainable clothes than their peers with lower incomes, it is projected that higher incomes will be connected with stronger purchasing power (Tey et al., 2018). According to a study, 34% of the global population is willing to pay a premium price for sustainable products or services in retail stores (Bernol, 2024). A broader trend towards mindful consumption is reflected when consumers are interested in investing a higher price in sustainable goods. Consumers are choosing minimally, items with higher quality that they can wear for longer periods, rather than frequently paying cheaper prices for fast, disposable fashion.

As a result, a sustainability-driven consumer base has been created past few years, which forces the current retail scenario to evolve as a landscape of consumers who prioritise eco-friendly clothing goods. Nowadays, online purchasing and entertainment platforms enable every brand to share their sustainability goals and information about their sustainability practices such as a transparent supply chain, from material sourcing to eco-friendly production process. E-commerce sites and applications are implanting features like virtual fitting rooms to reduce return rates and thus lower the carbon footprints associated with shipping. For people to have more participation in a circular economy, there is a rise in the purchase of resale and rental garments. By designing these useful changes, retailers are playing a crucial role in driving the fashion industry towards a more sustainable future and responding to the demands of the consumer as well. The company will not worry about its usual customer base and focus only on environmentally concerned consumers when the added utility from the sustainable items utilised by these consumers is very high (Yang & Dong, 2017).

6.7 Future Scope

The future of sustainable apparel retailing is bright and multifaceted, driven by several converging factors that promise to transform the industry. Technological advancements, such as AI, blockchain, and advanced textile recycling technologies, will play a crucial role in enhancing supply chain transparency, optimising inventory management, and enabling closed-loop systems where old garments are efficiently transformed into new fabrics. Innovative business models, including product-as-a-service, resale marketplaces, and on-demand manufacturing, will reduce waste and promote the reuse and longevity of garments, moving away from the traditional linear consumption model.

Consumer behaviour is also shifting, with increased awareness and demand for sustainable products. This will push retailers to provide greater transparency and authenticity, supported by clear, verifiable information about their sustainability efforts. Retailers will invest more in educating consumers about sustainable practices, encouraging responsible purchasing, and promoting garment care and repair to extend the life of products. This shift will be strengthened by stricter regulations and policies from governments' worldwide, recycling, waste reduction, and the use of sustainable materials, alongside incentives for eco-friendly innovations. Sustainable materials and fabric innovations, such as bio-based and smart textiles, will enhance product longevity and utility while sourcing from regenerative agricultural practices will improve environmental outcomes. Local and small-scale production, combined with collaborative consumption models like clothing swaps and sharing platforms, will reduce transportation emissions and support local economies. Retail innovations will also emerge, with eco-friendly store designs incorporating energy-efficient lighting and materials, and the rise of digital and virtual shopping experiences, including virtual try-ons and augmented reality (AR) fitting rooms, reducing the need for physical inventory and returns. Comprehensive sustainability metrics and reporting standards will become the norm, allowing retailers to measure and communicate their environmental and social impact effectively.

Global collaboration and cultural shifts towards valuing quality over quantity, driven by influential figures and movements advocating for sustainability, will support the growth of sustainable fashion. Increased investment in sustainable fashion startups and technologies will fuel further innovations and the adoption of best practices across the industry. By embracing these changes, the apparel retail sector can move towards a more sustainable, ethical, and environmentally conscious future, meeting the evolving demands of consumers and contributing to a healthier planet.

6.8 Conclusion

In conclusion, this chapter highlights how the industry transforms according to the sustainable initiatives taken towards the apparel retail sector. As there is an enormous imperative for sustainability in apparel retail, there are many strategies followed by the current clothing brands to sustain themselves. Green retail practices such as circular fashion and consumer education and engagement are improved to follow-up on sustainable goals within and outside the company. Sustainable raw materials like organic cotton and natural and recycled synthetic fibres are utilised to heavily promote environmental friendliness and eliminate resource inefficiency. As a comprehensive approach towards green practices, sustainable retail practices such as sustainable store design and organic packaging contribute to achieving sustainable apparel retail practices. Even though the production process of apparel goods comes before the retail sector, the eco-friendly manufacturing of such goods largely influences an ideal and sustainable apparel retail practice. Therefore, the eradication of ecological footprints in apparel manufacturing and retail is done by implementing eco-friendly manufacturing processes like the removal of non-toxic chemicals and unethical production techniques. Important parameters such as brand reputation, customer loyalty, and profitability are well enhanced while maintaining successful strategies for a sustainable retail outlet. Thus it not only helps for the growth of sustainability outside the firm but always checks the growth of the company as per its visions and missions to be ideally a profitable business. Effective collaboration and communication are necessary for a sustainable relationship between the retailer and supplier as it is pivotal to make the apparel retail industry consumer-friendly. The growing demand for a transparent supply chain and increasing awareness about environmental issues among consumers are influencing consumer behaviour towards the retail transition to a much greater extent. The apparel retail industry's dedication to sustainability serves as a model for other industries, proving that both financial gain and environmental responsibility are compatible. The clothing retail sector can make a substantial contribution to a more ethical and sustainable future by prioritising and consistently innovating in sustainable practices.

References

Aydin, N., Seker, S., Deveci, M., Ding, W., & Delen, D. (2023). A linear programming-based QFD methodology under a fuzzy environment to develop sustainable policies in the apparel retailing industry. *Journal of Cleaner Production, 387*, 135887.

Bernol, G. T. (2024). *Your customers are willing to pay more for sustainability, the motivations of the new consumer* [Online]. Available from https://bcome.biz/blog/your-customers-are-willing-to-pay-more-for-sustainability-the-motivations-of-the-new-consumer/. Accessed 20 June 2024.

Dabija, D. C., & Băbuț, R. (2019). Enhancing apparel store patronage through retailers' attributes and sustainability. A generational approach. *Sustainability, 11*(17), 4532.

Diversitech (2022). *A guide to sustainable retail industry 2024: Retail trends & consumer behavior* [Online]. Available from https://www.diversitech-global.com/post/guide-to-sustainable-retail-industry#:~:text=Not. Accessed 20 June 2024.

Fulton, K., & Lee, S. E. (2012). An exploratory study of sustainable apparel retailers on the internet. *International Journal of Electronic Marketing and Retailing, 5*(2), 95–109.

Gillivan, C. (2024). *Supplier relationships—How do they impact retail?* [Online]. Available from https://connorgillivan.com/supplier-relationships-affect-retail/#:~:text=Supplier. Accessed 20 June 2024.

Islam, S. (2021). Waste management strategies in fashion and textiles industry: Challenges are in governance, materials culture and design-centric. In *Waste management in the fashion and textile industries* (pp. 275–293). Woodhead Publishing.

Linkedin (2023). *Consumer behaviour changes and trends in retail by Retail Consulting & Advisory Services Ltd* [Online]. Available from https://www.linkedin.com/pulse/consumer-behaviour-changes-trends-retail/. Accessed 20 June 2024.

Marguerite LeRolland (2023). *Why integrating sustainability in fashion retail is becoming a must* [Online]. Available from https://www.euromonitor.com/article/why-integrating-sustainability-in-fashion-retail-is-becoming-a-must. Accessed 20 June 2024.

Medium (2018). *Top 5 environmental sustainability issues in retail by RCC* [Online]. Available from https://medium.com/retailcrc/top-5-environmental-sustainability-issues-in-retail-815aa0310779. Accessed 20 June 2024.

Mercier, P., & Battle, S. (2012). *Retailer-supplier collaboration in supply chain* [Online]. Available from https://www.bcg.com/publications/2012/retail-consumer-products-retailer-supplier-collaboration-supply-chain. Accessed 20 June 2024.

Muthu, S. S. (Ed.). (2014). *Roadmap to sustainable textiles and clothing: Eco-friendly raw materials, technologies, and processing methods*. Springer.

Rathinamoorthy, R. (2019). Circular fashion. In *Circular economy in textiles and apparel* (pp. 13–48). Woodhead Publishing.
Rausch, T. M., Baier, D., & Wening, S. (2021). Does sustainability really matter to consumers? Assessing the importance of online shop and apparel product attributes. *Journal of Retailing and Consumer Services, 63*, 102681.
Rodrigue, J. P., Slack, B., & Comtois, C. (2017). Green logistics. In *Handbook of logistics and supply-chain management* (Vol. 2, pp. 339–350). Emerald Group Publishing Limited.
Sagar, S. (2024). The impact of digital transformation on retail management and consumer behavior. *IOSR Journal of Business and Management (IOSR-JBM) e-ISSN, 26*, 06–14.
Shen, B., Li, Q., Dong, C., & Perry, P. (2017). Sustainability issues in textile and apparel supply chains. *Sustainability, 9*(9), 1592.
Tey, Y. S., Brindal, M., & Dibba, H. (2018). Factors influencing willingness to pay for sustainable apparel: A literature review. *Journal of Global Fashion Marketing, 9*(2), 129–147.
Timotius, E., & Octavius, G. S. (2021). Global changing of consumer behavior to retail distribution due to pandemic of COVID-19: A systematic review. *Journal of Distribution Science, 19*(11), 69–80.
Vadakkepatt, G. G., Winterich, K. P., Mittal, V., Zinn, W., Beitelspacher, L., Aloysius, J., Ginger, J., & Reilman, J. (2021). Sustainable retailing. *Journal of Retailing, 97*(1), 62–80.
Yang, L., & Dong, S. (2017). Sustainable product strategy in apparel industry with consumer behavior consideration. *Sustainability, 9*(6), 920.
Yang, S., Song, Y., & Tong, S. (2017). Sustainable retailing in the fashion industry: A systematic literature review. *Sustainability, 9*(7), 1266.

Chapter 7
Sustainable Reverse Logistics Practices in Apparel Retailing

R. Surjit, V. Pirnavi, S. Varshika, and S. Vasundhara Devi

7.1 Introduction

Sustainability has become a top priority in the fast-paced fashion world, changing the expectations of both businesses and consumers. Fashion companies are increasingly paying attention to the entire life cycle of their products, including the afterlife of the products and their disposal. There has been more concentration earlier on ensuring a smooth supply chain where the material reaches the consumer after passing through all the stakeholders of the supply chain. However, in today's context, there is also a greater need to look at the reverse flow of materials from the consumer to the manufacturer or the disposer. This reverse flow of material is required highly for various reasons like defects, servicing, end of use, end of life, etc. This reverse flow is known as Reverse Logistics. It can be understood better by the following definition. "It is the process of planning, implementing, and controlling the efficient, cost-effective flow of raw materials, in-process inventory, finished goods and related information from the point of consumption to the point of origin for the purpose of recapturing value or proper disposal" (Dowlatshahi, 2000; Srivastava, 2008; Stock, 1998). Reverse logistics has become very important due to the amount of material moved. A recent report by Skyquest (2023) valued the global reverse logistics market size at USD 946.63 and USD 1065.91 billion in 2022 and estimated it to grow to USD 2754.39 billion by 2031. It is expected to have a CAGR of 12.6% during 2024–2031. This indicates the magnitude of the business and the resources required for handling the same. It also requires to understand how it can be handled

R. Surjit (✉) · V. Pirnavi · S. Varshika · S. Vasundhara Devi
Department of Fashion Technology, PSG College of Technology, Coimbatore, Tamil Nadu, India
e-mail: rst.fashion@psgtech.ac.in; 22H228@psgtech.ac.in; 22H243@psgtech.ac.in; 22H246@psgtech.ac.in

S. S. Muthu (ed.), *Sustainable Apparel Retailing*, Sustainable Textiles: Production, Processing, Manufacturing & Chemistry,
https://doi.org/10.1007/978-3-031-83875-0_7

sustainably. The various types of services in reverse logistics are returns management, recycling, refurbishment, disposal, and remanufacturing. Reverse logistics has become a point of contention after e-commerce business growth as many online retailers deliver the products to the customer and then the customer returns them if there are any issues with them. These returns have to be handled in the most optimised way to avoid environmental damage and reduce the carbon footprint. The strategic importance of reverse logistics as an enabler of sustainable performance has attracted much attention. In the past, logistics in the fashion industry focused on efficient delivery of materials to the customer whereas today, it also looks at the post-consumer delivery handling and servicing. Reverse logistics includes operations such as recycling, repair, exchange and return. One of the main benefits is that it extends the life of products, which reduces waste. Returned items can be reused, resold or refurbished instead of ending up in a landfill, reducing the environmental impact of fashion consumption. However, integrating reverse logistics into fashion supply chains has its challenges. Returns can be difficult and expensive to manage; it requires careful planning, investment in infrastructure and technology, and collaboration. Fashion brands use creative partnerships and solutions to solve these problems. By focusing on sustainability, recyclability and repairability, brands can minimise waste. This is where many reverse logistics companies come in where they are specialised in implementing efficient reverse logistics systems that can improve customer satisfaction and brand loyalty. By offering free returns and transparent communication throughout the process, fashion companies can build consumer trust and encourage repeat purchases. Integrating reverse logistics into business strategies is not only a necessity in the fashion industry, but it is also a brand value enhancer as customer satisfaction is of utmost importance in this business. An effort has been made in this chapter to explain the concept of reverse logistics, the process of returns, challenges in reverse logistics, sustainable practices followed in reverse logistics, various brands' focus on reverse logistics, the role of reverse logistics companies and the emerging opportunities in reverse logistics.

7.2 Reverse Logistics

Before one understands reverse logistics, a small overview of logistics and its role is to be discussed. Logistics is the process of moving the materials from the manufacturer to the customer. The supply chain management involves various stakeholders starting from raw material producers and ending with customers with manufacturers, distributors, wholesalers, and retailers in the middle. The supply chain management will ensure the proper flow of goods, information, and money so that the customer gets the intended product for usage. In the initial days, logistics was referred to as supply chain management and over some time, the usage of the term logistics was restricted to the transportation and distribution of the materials alone under the broad ambit of supply chain management (Shah, 2009). The logistics framework encompasses several key elements that ensure alignment with customer demands, such as managing prices and quantities. Quality assurance ensures the accuracy of procurement and delivery, while processing regulates and controls the time needed to move products from supply to issuance. Logistics is essential for maintaining high productivity, as it reduces the time required to acquire necessary products for manufacturers or those delivering goods and services to end consumers. While small businesses often focus on designing and manufacturing products to meet customer needs, logistics ensures these products reach customers. Efficient logistics is crucial for profitability and overall business success. Effective management of resources for timely delivery and handling of materials can significantly impact a business. On the customer side, timely manufacturing and delivery of products are essential for maintaining customer satisfaction, which directly affects the company's profitability and long-term viability. The logistics process involves the process of delivering the right product at the right time, the right cost, the right quality, the right quantity, and the right place to the right customer.

The above same logistic process applies to reverse logistics too as it involves moving the product from the customer or any of the stakeholders in the supply chain in the reverse direction due to requirements like repairs, refurbishment, quality issues, etc. The process by which consumers return goods to the manufacturer or distributor for repair, refund, exchange, repackaging, recycling, and other purposes is called reverse logistics. The reverse movement of materials is sometimes inevitable due to the complex nature of businesses today and the huge volume of materials sold to the customer for their needs, and desires. For example, if a customer got damaged apparel delivered for an online order and wants to return the product, the reverse logistics process comes into action where the e-commerce platform initiates the return process. The material has to be picked up by the courier company and returned to the manufacturer or the seller so that the new material can be delivered to the customer. Now the material returned has to be handled by the manufacturer by sending it for reprocessing where the defects are corrected and then the material is repacked for sending to the next customer when there is an order for the same. If in case, the damage is not rectifiable, then the reverse logistics process will still be in existence as the disposal of the product has to be decided and the product has to be moved for the same. If the product is capable of being recycled, then it has to be moved for recycling instead of disposal. All these movements are part of the reverse logistics process and so all these have to be done at the right time, right place, right material, and at the right cost thereby signifying the importance of the process and its requirements. One of the main strategies that companies use to differentiate themselves from their competitors is reverse logistics. Companies using reverse logistics have well-organised systems. For the customer, this means the ability to satisfy their demands, build customer loyalty, and make it a challenge for competitors to copy clear services that prioritise value creation and the environment. The need for reverse logistics is mainly driven by two factors: first, companies begin to see the economic benefits of effective recovery management, and second, changes in the regulatory environment require manufacturers to take full responsibility for their products throughout their life cycle. Hence it plays a significant role in our fashion field.

7.2.1 Closed-Loop Supply Chain

It is imperative to understand the open-loop supply chain first before analysing a closed-loop supply chain. For a long time, open-loop supply chains were being followed where the used products were sent for recovery to a different firm other than the original manufacturer. For example, a shoe manufactured by a leading brand is sent to a waste management company after its use to extract usable material content from the shoes. In contrast to this, a closed-loop supply chain is in operation now where the products are sent to the manufacturer after its usage for recovery of materials from it. A closed-loop supply chain has a lot of advantages. One of those is that the manufacturer becomes responsible during manufacturing as the return materials have to be processed by him/her and if they have any material that is difficult to recycle, they will find it more cumbersome and also lead to more cost and effort utilisation. The closed-loop supply chain encompasses the reverse logistics process. As discussed earlier, reverse logistics is the process of returning a delivered product to the seller or manufacturer, following the same supply chain path but in reverse. This process starts with, sometimes passing through warehouses or distribution centres. Figure 7.1 explains the reverse logistics process in detail. The forward and reverse logistics when combined bring a closed-loop supply chain into place. Closed-loop supply chain management involves the development of a system to maximise the value of the entire lifecycle of a product including the value recovery (Guide Jr & Van Wassenhove, 2009). The closed-loop supply chain shown in Fig. 7.1. explains the forward and reverse process in the closed-loop supply chain. The forward logistics process is denoted by a solid line and the reverse logistics by dashes. The forward logistics process is comparatively easier as the material is moved from the raw material manufacturer to the manufacturer and then to the distribution and customer. However, reverse logistics involves return evaluation from

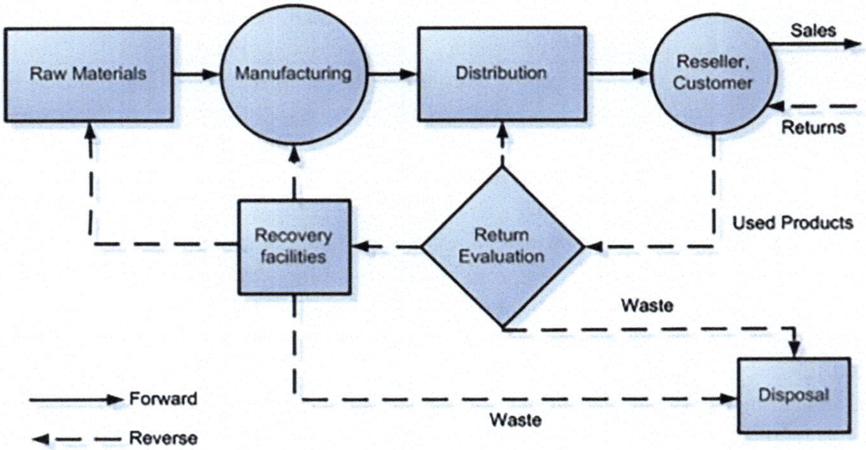

Fig. 7.1 Closed-loop supply chain (Govindan et al., 2015; Tonanont, 2009)

any of the forward logistics stages. After evaluation, it is decided to either move it as waste for disposal or send it to the recovery facility to rework the material and then be sent to the distribution again. Even during the recovery, waste is generated because of faulty materials or broken parts and is sent for disposal.

Various activities are part of reverse logistics like repairing, remanufacturing, and refurbishing products, disposing of obsolete items, processing returned products, reusing packaging material or recycling them, disposal of hazardous materials, and recovering assets. It has a vital role to play in reducing the cost of reprocessing products as it should not be higher than the new product thereby ensuring profit. It also involves the flow of information other than the flow of the product as all the stakeholders in the supply chain should be aware of the return and the way it is handled. Reverse logistics is crucial for handling failed deliveries, damaged goods, and product returns in both B2C and B2B transactions. Small businesses often outsource their reverse logistics needs to specialised third-party logistics (3PL) companies, which facilitate the return of products from their origin to their destination. Due to its strong correlation with consumer satisfaction, reverse logistics is an essential component of the e-commerce supply chain. One of the most important aspects of reverse logistics is returns. Businesses must establish a proficient procedure for managing product returns. It involves putting in place a system to keep track of repairs, swaps, and refunds. It also entails making sure that any returned goods are properly inspected and handled. Customers ordering online could receive items that are not what they expected, such as the wrong size or colour. Reverse logistics is necessary in these situations when customers swap or return goods. It is imperative to understand the various types of returns. They are detailed below:

- Warranty returns: These are the returns that arise because of defects in the product that are in the warranty period.
- Manufacturing returns: These returns arise when the product has issues during manufacturing and require to be disposed of or used for some other product due to unsatisfactory quality or the leftover materials are returned.
- Service returns: These are the returns that arise for servicing the products that were used by the customer.
- Commercial returns: Such kinds of returns arise due to contracts between the manufacturer and seller wherein the manufacturer agrees to take back the obsolete products after a certain time.
- End-of-use returns: Once the intended end use is achieved, the customer returns the product and then the manufacturer can try to recycle or reuse the product.
- End-of-life returns: These returns result when the product lifespan is completed and it needs to be disposed of properly.

7.2.2 Reverse Logistics in Apparel and Fashion Supply Chain

Reverse logistics has an integral part to play in the apparel and fashion supply chain as it takes care of returning the product from the end consumer to the manufacturer or merchant. This process involves various planning and control strategies, and some companies may choose to outsource it to specialised providers enabling the return of goods for processes like recycling, refurbishing, or returns. The process starts with the end user and flows backwards through the supply chain, involving distributors, manufacturers, and other stakeholders. Effective return management is crucial to a company's profitability, as it involves overseeing and controlling the reverse flow of goods from customers back to the seller or manufacturer.

7.2.2.1 Process of Return

The return process is initiated when a consumer requests to return a product, triggering a series of steps. First, a return authorisation is issued, and the product's condition is assessed. Next, a return shipment is scheduled, and the refund is approved. If the product is faulty, a replacement is arranged. This process requires efficient coordination to ensure a smooth and satisfactory experience for the consumer. Effective return management involves streamlined communication, accurate product evaluation, and timely resolution, ultimately building trust and loyalty with customers.

Upon receiving a returned product at the facility or centralised processing centre, a thorough inspection is conducted to determine its condition and categorise it accordingly. Ideally, with optimised reverse logistics, the product's destination should be predetermined before arrival. Next, the products are sorted into appropriate disposition categories, including:

1. Repair and refurbish: Fixing the product to restore its original functionality. By implementing repair and refurbishment processes, businesses can significantly reduce the losses incurred because of the disposal of the product without usage. This approach also offers cost savings and enhanced supply chain sustainability. Moreover, it provides an opportunity to enhance customer satisfaction by ensuring high-quality and long-lasting products. By adopting repair strategies, businesses can reduce waste, improve efficiency, and promote sustainability throughout their operations.
2. Resell as new: The product has no defects and it is returned because the customer did not like it because of the colour or size or specifications, etc. In this case, the product is again sold as new.
3. Resell as returned: The product may have minor imperfections but can still be resold at a lower price.
4. Recycle: Product materials can be recycled and transformed into new products. When a product reaches its end-of-life (EOL) or repair is no longer viable, retailers can explore responsible disposal options like recycling or donation. By

choosing eco-friendly disposal methods over landfilling, brands can appeal to the growing number of consumers who prioritise environmental sustainability in their purchasing decisions, enhancing the brand's reputation, and contributing to a more circular economy.

5. Scrap: The product cannot be repaired or resold and must be disposed of.

Accurate sorting and disposition decisions are crucial in reverse logistics to maximise product value recovery and minimise waste. By returning products to their original supplier, businesses can implement sustainable practices such as reselling, remanufacturing,

7.2.2.2 Cost of Reverse Logistics

Understanding the costs associated with reverse logistics is required. The movement of materials in reverse direction involves a lot of costs. The various costs associated are the cost of holding the material, the cost of returning the material, the cost of inspection of products, the cost of rework, and the cost of defective products. All these costs are incurred at the manufacturer's end and except the rework cost, other costs are incurred by the retailer, returning centre and the warehouses handling the products. The cost of defective products may occur at all the departments because the defect could have arisen because of any of the stakeholders in the supply chain like the warehouse department, the transport department, or the retailer end. Hence, this cost has to be borne by the department which caused the defect.

Costs at Retailer End

At the retailer's end, when a product is returned, significant costs are incurred. The retailer has to hold the material till it is sent to the returning centre or the corrected/refurbished/repaired product is sent back to the customer. This holding cost depends on the number of pieces that have to be returned and the time for holding them before they are sent to the returning centres or warehouses. The holding cost can be reduced if the materials are transferred to the returning centre rapidly but since it involves transportation costs, it can only be sent when there are more defective pieces or multiple retailers have defective pieces to be sent. It is a trade-off between the holding cost and transportation cost always based on the situation. The transportation cost for the retailer to send the material back to the returning centre or warehouse could be significant if the number of pieces to be sent back is much less thereby increasing the unit transportation cost. The transportation cost is dependent on the number of units and the unit cost of transportation (Singh et al., 2016). These costs are incurred by the retailer or by the manufacturer based on the agreement they have executed. It is essential to look at the costs at the retailer's end and try to reduce them so that it can lead to significant savings. Also, unnecessary transportation leads to pollution and wastage of resources.

Costs at Returning Centre

There are significant costs involved in handling the returned material at the return centre. It is similar to the costs incurred at the retailer's end. The returning centre can be a commonplace to collect all the return items of a particular manufacturer and later it is sent to the warehouse. In some cases, the returning centres and warehouses can be the same. In cases, where there are more returns and also warehouses are at very distant locations, then it is essential to establish such return centres to first collect the products and then send them to the warehouse. This centre also incurs holding costs till the products are sent to the warehouse and also, transportation costs to send the material to the warehouse. If the returning centre and warehouse are the same, then there is a cost of inspection of products too.

Costs at Warehouse

The warehouse is the place where the product is inspected to confirm the reason for the return and to find the nature of the defect if any. Based on the inspection, it is decided whether the product should be sent to the manufacturing centre or sold in seconds or regarded as scrap to be disposed of. All these incur costs as sending the material to the manufacturer would involve transportation costs and selling as scrap or seconds would reduce the value of the product leading to a loss because of the product. Apart from this, there will be holding costs and inspection costs incurred in the warehouse. There will also be a cost of transportation for moving the material to the manufacturing centre or again to a retailer to sell it in seconds. There may not be costs associated if sold as scrap as the scrap dealer would pay an amount for the material which would generate revenue.

Costs at Manufacturing Place

The product is reworked in the manufacturing centre to correct the defects and then if it is workable, rework is done. After rework, the material is returned to the warehouse for the forward logistics process to work. If not workable or repairable, then the material is sent to the secondary retailers or markets to be sold in seconds or disposed of as scrap. In the manufacturing centre, there are costs of transportation to the warehouse, cost of reworking, cost of holding inventory, cost of transportation to the secondary market and also revenue from the material sent to the secondary market and as scraps.

The above-mentioned costs are incurred at the various departments in a supply chain when the reverse logistics works. The total cost of return will be the sum of all the costs mentioned above minus the revenue generated from seconds sales or scrap disposal. These costs discussed above give a clear picture of the proportion of costs involved in processing returns. There are significant costs incurred in reverse logistics and it is essential to minimise these costs as much as possible.

7.3 Challenges in Reverse Logistics

There are various challenges in the reverse logistics process. This is because the entire process involves many stakeholders like customers, the pick-up team, the returning centre, the warehouse team, the manufacturing centre, etc. There are significant time and costs involved in this process and hence, there are problems like the failed collection of returning items, reverse logistics planning and execution (Farokhi, 2011), financial costs, lack of information (Srivastava, 2008), competition, space availability, customer frauds, and increased carbon emissions.

7.3.1 Failed Collection

Failed collection can occur for a variety of reasons, such as the customer not being home, incorrect addresses, or discrepancies between the delivered product and the one returned. In these situations, the reverse logistics process is typically initiated by the delivery driver. This method minimises both storage and distribution costs by utilising existing delivery routes and personnel. When outsourcing reverse logistics services, it is essential to address whether additional charges will apply for handling failed collections. Since the driver is already scheduled to return to the distribution centre, it is important to determine if coming back without the goods incurs extra costs. Clear communication and understanding of these charges can help manage expenses and streamline the reverse logistics process effectively.

In another case, when an order arrives at the customer's location with damaged items, the delivery driver upon discovering that some or all items are damaged, the driver should identify and separate the damaged products from those that are intact. The damaged items should then be sent for further inspection. To facilitate a smooth return process, there should be a clear note stating which items were accepted by the customer and which were returned due to damage. This documentation helps ensure accurate processing and helps the retailer address the issue effectively. It also aids in managing inventory records and initiating any necessary claims or replacements. Implementing a standardised procedure for handling damaged returns can improve efficiency and customer satisfaction while minimising disruptions to the supply chain.

7.3.2 Reverse Logistics Planning and Execution

The planning and execution of reverse logistics involve a greater commitment of all personnel and a foolproof process to ensure how the returns are going to be processed. It is easy till the returns are lesser but when the number of returned items increases, it becomes difficult for the retailers or the manufacturers to plan effectively and efficiently as it involves a unified plan encompassing all the people involved in the process. There is also the challenge of inadequate infrastructure to deal with the returns. So it is required to create a proper plan on how to handle return items and also at the shortest possible time.

7.3.3 Financial Costs

As discussed in the above section, there are significant financial costs involved in reverse logistics and the challenge arises on how the costs can be managed and who in the supply chain will manage it. Since reverse logistics are significant, it would affect the revenue and profit of the company and so the retailers and manufacturers should be wary of the costs incurred. On the other hand, who will be responsible for the costs is another challenge that has to be decided mutually between the stakeholders. It is because, during the inspection, it may be said that the damage is not because of the manufacturer and it could be due to the warehousing personnel mishandling. For example, if a customer orders apparel and receives a damaged package, then the customer might not accept the package. Now, when the material is returned, the warehouse processing centre or the manufacturer may claim that the package has been damaged by the retailer or the delivery person and not by them. In this case, the person who will bear the costs becomes a question mark and a big challenge. Hence, there is a requirement for a standard operating procedure (SOP) in these cases to avoid conflicts.

7.3.4 Lack of Information

Supply chain management is the process of flow of information, funds and material. If there are stoppages in any of these flows, then it leads to supply chain disruption. Since reverse logistics is also an integral part of the supply chain, the above disruptions cause a problem in that too. In this case, it is the lack of information or improper information flow that would pose a serious threat to the process. There may be foolproof forward logistics process setup but the reverse logistic activities may not have been given due importance. Hence, there is a chance that the information might not be known to all concerned about the returns and ways to handle the returns. The customer might have been told about the pickup of the product within a specified time and if that information is not correctly passed on to the other people responsible for collection and further processing, it might disrupt the whole reverse logistics process. There should be a clear process to ensure the uninterrupted flow of information.

7.3.5 *Space Availability*

It is a big challenge as the returned items would occupy significant space in the warehouse and may lead to space insufficiency. Many times returns are not expected and the retailer or the manufacturer might not have planned the amount of returns and would not have allocated the space for keeping them. If the returns are significant, then there might not be enough space to keep them and it also involves the holding cost of the material. So space availability should be planned to avoid issues.

7.3.6 *Other Challenges*

There would be competition from other retailers, manufacturers or logistics companies in the handling of reverse logistics. So it is essential that a clear system is in place to handle reverse logistics and have a distinct advantage over the competitors. The other challenge can be customer fraud where the customer might have returned some other material or non-defective item, etc. Proper care should be taken to inspect the material when the customer returns it. Last but not least, the increased carbon emissions are a very big challenge when dealing with the returns as the returns processing involves considerable damage to the environment and causes pollution. So having a green supply chain and reverse logistics process is essential.

7.4 Sustainable Practices in Reverse Logistics

As mentioned above, a green reverse logistics process is always required especially in the handling of fashion materials. It is because, these materials cause considerable damage to the environment because of their material contents, logistics, etc. and then if those materials are returned, reprocessed, disposed of, etc., they might add to the burden of an increase in carbon footprint. Hence, it is required to implement sustainable practices in reverse logistics. The various sustainable practices that can be incorporated include having a closed-loop supply chain with enhanced reusing, repairing, refurbishing, and remanufacturing. Along with this, sustainability can be obtained by a circular economy, better product design, improved data analytics, vital collaborative strategies, technological innovation, and green logistics.

7.4.1 Closed-Loop Supply Chain

The closed-loop supply chain is the process that involves utilising the returned items more prospectively. The process involves the reuse of the returned items and also the recovery of the materials from the returned product. Major retailers try to refurbish the items and sell them to the customers so that the actual value of the product is realised. Such effective closed-loop supply chains can enhance the usage of manufactured materials and also lead to reduced waste and an improved environment. The closed-loop supply chain also involves concepts like reusing, repairing and remanufacturing where the maximum utility of the manufactured product is realised as the returns involve a lot of costs.

7.4.2 Circular Economy

Circular fashion derived from circular economy is a very important cog wheel of sustainability. The circular economy involves utilising the products to the fullest possible extent by finding new ways to use them and not leading to quick disposal. It involves processes that increase the lifespan of the product and that too when a material is returned, the circular fashion concept seeps in thereby creating a need to handle the product better to avoid wastage and pollution. So what is done in the circularity model is that when the product is returned, it is checked for defects or issues reported and then based on the assertion, the next step is decided. It could be reusing the product, upcycling the product or reselling the product as the process involves utilising the product for the maximum time possible before disposal. Also, there are various other models involved circularly like renting, swapping, etc. which may be used based on the brand's interests. However, every step should be taken to ensure the longevity of the product in some form before disposal.

7.4.3 Product Design

This is one area where every brand or manufacturer is concentrating on as the product design plays a key role in sustainability. A lot of thought process and brainstorming goes into product designing as brands now look for sustainable materials for manufacturing, sustainable production technologies and sustainable designing. Sustainable product designing is kept in mind as it would lead to the creation of a product that has easier end-of-life handling processes than the unsustainable ones thereby leading to an easier disposal process in the future. Having such product designs eases the process of return handling as the brands are aware of the return process and initiate them with ease when there is a return. It also leads to the effective recovery of the materials used in the product.

7.4.4 Data Analytics

In the current context of data being the big thing, more data helps in getting meaningful information to process the returns and handle reverse logistics effectively. The arena has moved from mere data collection to mining and analysis to extract meaningful information that can be used for sustainable reverse logistics handling. With the data collected from the forward and reverse logistics of a brand, one can apply various data analytics tools to improve various aspects of reverse logistics like information handling, environmental concerns, returns assessment, quality and quantity of returns, returns handling capabilities, personnel required for managing returns, and areas of improvement to name a few.

7.4.5 Collaborative Strategies

Collaborative strategies like vendor-managed inventory (VMI) and quick response (QR) have helped apparel retailers and brands to a great level to enhance customer service. The efficient consumer response (ECR) helps in handling customer issues effectively and leads to easier returns processing. The collaborative strategies include collaboration with all the stakeholders so that the best practices can be shared among all and lead to more improved and enhanced returns handling that can reduce costs and environmental impact. Such networks help in sharing the resources and expertise among all and also lead to more support on all fronts. It helps companies to leverage their abilities in the best possible way and also leads to the proper flow of information, funds and material in the reverse logistics flow.

7.4.6 Technological Innovation

Technological innovations are at the forefront now as there are a lot of new technologies that have come up that enable reverse logistics to function effectively. Every brand is making its online presence felt and hence deliveries are made through logistics providers. Such providers use enhanced technologies like QR codes, same-day deliveries, instant deliveries, immediate cashback, etc. Technologies like artificial intelligence (AI) have enabled businesses to grow at a very fast pace. The technology has enabled returns processing to be done in record time and it also assists in identifying the further processes that can be carried out with the returned items. There are standard operating procedures (SOPs) devised by such technological advancements that suggest the probable usage or process for the returned material.

7.4.7 Green Logistics

Green transportation or green logistics is a vital sustainable practice in the reverse logistics process. Walmart introduced a green supply chain system initially which involved utilising transportation facilities that cause minimal harm to the environment and protect nature. Such systems should be incorporated to ensure the material is transported in an environmentally friendly vehicle or follows the processes that result in a minimal carbon footprint. There is a need to look at alternative transportation methods available to carry the return materials so that the load on the environment is reduced. There can be usage of electric vehicles or vehicles with minimum emission for such purposes. Also, proper logistics planning can be done to transport the returns collectively and in a time-bound manner.

Such initiatives are not exhaustive and are only a few that can be used to improve the sustainability of the reverse logistics process in the apparel retail business. However, they give a good idea of what can be done to enhance sustainability in the reverse logistics arena and how the retailers and brands can work together to develop effective systems to handle returns with minimum ecological imbalance.

7.5 Apparel Brands Focus on Reverse Logistics

Reverse logistics is an integral process for apparel brands and as there is more online selling taking place, returns processing has become a focal point for the brands to look into. Returns and reverse logistics process is inevitable but it can be carried out with more ease and reduced load on the environment. Many brands are taking different routes to handle the reverse logistics and returns processing. In this section, the brand's focus on reverse logistics is discussed.

7.5.1 H&M

H&M has dedicated itself to sustainable development for nearly three decades. Their philosophy is to offer customers a balance of fashion, quality, price, and sustainability. They believe sustainable fashion should be accessible to all, not just a select few. While they have made significant progress, there is still much work to be done. The industry needs to reduce its environmental impact, improve social responsibility, and build a more sustainable fashion system. They are committed to transparency and collaboration to achieve these goals. Their focus areas include leading change, measuring innovation, promoting transparency, and advancing recycling and climate initiatives. They aim to operate within planetary boundaries by 2040 and have a positive impact on biodiversity and people throughout our value chain. They also support and promote inclusion and diversity in all our work. One key initiative is their reverse logistics programme, inspired by H&M's successful approach. They collect used clothes from all brands, repair, reuse, remanufacture, and recycle them to create a circular fashion system. Their goal is to encourage customers to donate their old clothes, reducing waste and supporting sustainability. The H&M Reverse Logistics process involves clothing collection, sorting, and processing. They have collected over 155,000 tons of clothing since 2013, with items either reused, recycled, or remanufactured into new products. They are working to strengthen their collection and recycling processes, collaborating with policymakers to become a more circular business. By embracing sustainable development and reverse logistics, H&M believes they can reduce waste, support the environment, and promote a more responsible fashion industry.

7.5.2 *Myntra*

Myntra is a popular online retailer in India dealing with apparel products. They have built an effective reverse logistics system to handle and reduce returns. Their various attempts to decrease the returns are discussed here. They have a preventive methodology where they keep a tab on the order return rate per user. This is based on the number of orders placed and returned by the customer. This rate is compared with the average returns rate and if the order return rate of a particular customer is high, then a few measures like removing the rewards benefits, adding a non-refundable convenience fee, etc. are taken. Also, Myntra avoids returns by getting complete anthropometric details of the customer along with their preferences to suggest the best-fit products for them to avoid returns. They also use AI tools to help customers select the right products. They have an effective reverse logistics process empowered by LogiNext (2024) that optimises return delivery routes and provides an extensive intra-city and inter-city collection network. Also, provides delivery tracking in real-time, manages a large fleet of vehicles, captures customer location data in real-time and optimises the last mile collection along with huge savings on reverse logistics costs.

7.5.3 *Ajio*

Ajio, a famous online apparel brand of Reliance Group, has an effective reverse logistics process handled by its reverse logistics partners. Addverb (2024) is a partner of theirs and they specialise in reverse logistics. They handle the reverse process sustainably and effectively by following an effective process. Their reverse logistics process has significant advantages. They are

- Improved accuracy and efficiency as their systems identify the items and segregate them for reuse, repair, recycle, etc.

- Streamlined return process as warehouse automation leads to continuous processing of returns.
- Quicker and efficient returns to the customer making them satisfied by keeping them updated on the return process after collecting the material from them.
- Having accurate data on the inventory available for effective repair or reprocessing.
- Significant cost reduction by automating warehouse returns processing.
- Reducing human error by automation and thereby reducing customer complaints arising because of them.
- Sustainable process as the environmental impact of handling products including disposal is reduced due to effective promotion of recycling, reusing, and disposing.
- Enhanced data analytics, scalability of operations, and transparency in the process lead to effective returns handling.

7.5.4 Amazon

The discussion would not end without mentioning the reverse logistics process of one of the most customer-centric companies and that is Amazon. They deal with all kinds of products including apparel. Their reverse logistics process is more sophisticated and dependent on the latest technologies. The returns are handled separately and fulfilment centres are not used for the same. The process involves using advanced algorithms and software to decide the optimum travel of the return product for repair or refurbishing. They also decide whether the product needs to be sent back to the seller or disposed of based on the condition, demand, and inventory availability of the product. Amazon also has a gift card concept to reduce the burden of returns as the money is used by the customer in future to buy some products from Amazon. There is an integrated network to share information with sellers so that they can decide which products they want to receive back and which can be disposed of according to the value of the product and the demand. They also resell the items quickly to recover the value. Their automation and data-based processes help in effective returns management thereby improving the sustainability of the whole process.

The above activities of brands are just a few excerpts from their various works in handling reverse logistics effectively and efficiently. The apparel brands are coming out with new strategies as we read this on how to manage reverse logistics and returns more profoundly thereby conserving the environment and reducing the costs associated with the whole process.

7.6 Role of Reverse Logistics Companies

Reverse logistics companies provide a comprehensive range of services that align with the five R's of reverse logistics. These services include:

1. Returns management: Retrieving returned products.
2. Inspection: Checking products for damage or defects.
3. Repackaging and relabelling: Preparing products for resale according to seller requirements.
4. Replacement: Exchanging defective products with new ones during the return process, saving time and money.
5. Refunds: Processing refunds and providing omnichannel support to vendors and customers across all platforms.
6. Status updates: Keeping customers informed about pickup or exchange status, ensuring transparency and communication throughout the process.

By offering these services, reverse logistics companies facilitate efficient and cost-effective returns management, enhancing the overall customer experience and supporting businesses in managing their reverse logistics operations. By partnering with specialised reverse logistics companies, businesses can reap a range of benefits that enhance their overall operations. These advantages include:

1. Improved customer satisfaction: Efficient returns processing leads to increased customer loyalty and positive feedback.
2. Reduced damage costs: Expert handling minimises damage and prevents losses.
3. Simplified product reuse: Returned items are managed and prepared for resale or recycling, streamlining the process.
4. Enhanced sustainability: Environmentally responsible disposal and recycling options boost the supply chain's value and eco-friendliness.
5. Actionable insights: Reverse logistics experts provide valuable data to optimise operations and refine the returns process.
6. Optimised delivery: Advanced technology and route planning enable efficient collection and delivery of returned goods, saving time and resources.

By leveraging these benefits, businesses can strengthen a more sustainable and efficient supply chain. Also, the company can be selected based on the following essential factors to ensure a successful collaboration:

1. Sector-specific expertise: Opt for a company with a proven track record in your industry, understanding the unique challenges of your product returns.
2. Comprehensive service offerings: Ensure the company provides the necessary services, including return types, geographical coverage, and value-added services like repackaging and quality control.
3. Technological prowess: Select a company that leverages advanced technology for efficient tracking, data analysis, and seamless integration with your e-commerce platform or systems.

4. Extensive network coverage: Consider a company with a robust network that can handle returns from your target locations, catering to your regional or national customer base.
5. Flexible pricing: Compare pricing models and choose a company with a flexible structure that aligns with your return volumes and budget.
6. Commitment to sustainability: Evaluate the company's dedication to eco-friendly practices, such as responsible product disposal and recycling, if these align with the company's business values.

There are various reverse logistics providers. It is better to have the providers work on them as they are equipped with better practices. There are third-party logistics (3PL) and fourth-party logistics providers (4PL) who do effective work in handling reverse logistics. The 3PL providers handle the entire work of handling the flow of material, information and funds in a closed-loop supply chain and that gives enough room for handling reverse logistics. The 4PL companies work more by incorporating concepts like reinvention, transformation, and execution where they take care of all the work other than manufacturing and sometimes also provide consulting for all the activities including manufacturing. Such providers have very efficient systems to handle returns effectively. The 4PL companies also devise sustainable strategies that can be used by their partners to improve their activities and also result in mutual growth for all. These logistics providers enhance sustainable practices by following a green supply chain and various other initiatives that lead to a reduced environmental footprint.

7.7 Emerging Opportunities in Reverse Logistics

To enhance reverse logistics and returns management, online retailers can follow five best practices:

1. Streamline the returns process by providing clear instructions, prepaid shipping labels, and regular updates on return status.
2. Automate returns management using software that generates shipping labels, tracks returns, and issues refunds, reducing processing time and errors.
3. Prioritise customer experience by making returns easy to initiate and providing clear instructions, updates, and quick refunds or exchanges.
4. Analyse return data to identify trends and patterns, optimising processes and reducing costs by addressing quality control issues or improving product descriptions.
5. Efficiently manage returned products by checking their condition, reselling or finding alternative uses, and reducing waste through reuse or recycling.

The e-commerce boom has forced retailers to re-evaluate their strategies, recognising the imperative need for effective reverse logistics management to maintain competitiveness, reduce costs, and enhance customer satisfaction. By prioritising reverse logistics, retailers can turn a potential liability into a strategic advantage.

In the pursuit of green logistics and sustainability, many retailers, including prominent US retailer Best Buy, are integrating on-demand reverse logistics services to manage the disposal and recycling of outdated electronic appliances and other household items. This initiative reflects a broader commitment to environmentally friendly practices. To enhance reverse logistics efficiency and promote sustainable operations, these retailers coordinate the collection of old products for recycling alongside the delivery of new orders. By synchronising the pick-up of used items with the delivery of fresh purchases, retailers can streamline logistics, reduce transportation costs and minimise the carbon footprint associated with separate trips. This integrated approach not only supports recycling efforts but also optimises delivery routes and improves overall operational efficiency.

7.8 Conclusion

Reverse logistics is a very effective and most required process as it handles the returned materials in the best possible way. The returns are to be effectively managed to enhance the closed-loop supply chain process. Such returns must be handled sustainably so that the environment is protected and life on Earth becomes easier for future generations. The reverse logistics process and the costs involved in reverse logistics show the complexities of the process and open many challenges in it including space requirements, failed collections, etc. thereby stressing the need for careful planning. Various sustainable practices can be incorporated to enhance the reverse logistics process like green logistics, data analytics, etc. to make them deal with returns effectively and mitigate the costs involved in it. Such practices will help apparel retailers in a great way to resort to sustainable branding practices. The apparel brand's initiatives are also discussed in this chapter and that signifies their responsibility towards protecting the environment. The role of reverse logistics and emerging opportunities are also discussed in this chapter. It is understood that proper planning of reverse logistics activities will lead to a sustainable future as there are significant contributions from this process to environmental sustainability.

References

Addverb. (2024). *Warehouse automation* [Online]. Available from https://addverb.com/warehouse-automation-can-change-reverse-logistics-game/. Accessed on 20 June 2024.

Dowlatshahi, S. (2000). Developing a theory of reverse logistics. *Interfaces, 30*(3), 143–155.

Farokhi, S. (2011). A dual model for describing of reverse logistics inventory systems. *International Journal of Modeling and Optimization, 1*(2), 174.

Govindan, K., Soleimani, H., & Kannan, D. (2015). Reverse logistics and closed-loop supply chain: A comprehensive review to explore the future. *European Journal of Operational Research, 240*(3), 603–626.

Guide Jr, V. D. R., & Van Wassenhove, L. N. (2009). The evolution of closed-loop supply chain research. *Operations Research, 57*(1), 10–18.

LogiNext. (2024). *Myntra's cost cutting journey* [Online]. Available from https://www.loginextsolutions.com/blog/myntra-last-mile-delivery/. Accessed on 20 June 2024.

Shah, J. (2009). *Supply chain management—Text and cases*. Pearson Education.

Singh, R. K., Acharya, P., & Taneja, D. (2016). Impact of reverse logistics on apparel industry. *International Journal of Services and Operations Management, 25*(1), 80–98.

Skyquest. (2023). *Reverse logistics market insights* [Online]. Available from https://www.skyquestt.com/report/reverse-logistics-market#:~:text=Global%20Reverse%20Logistics%20Market%20size,period%20(2024%2D2031). Accessed on 21 June 2024.

Srivastava, S. K. (2008). Network design for reverse logistics. *Omega, 36*(4), 535–548.

Stock, J. R. (1998). *Development and implementation of reverse logistics programs*. In Annual conference proceedings, council of logistics management.

Tonanont, A. (2009). *Performance evaluation in reverse logistics with data envelopment analysis*. In Proceedings of the 2008 industrial engineering research conference.

Chapter 8
Facilitating Sustainable Conversations: Towards a Human-Like Chatbot for the Fashion Industry

Vanissa Wanick, Eirini Bazaki, A. R. D. B. Landim, R. J. R. dos Santos, G. Paiva Medeiros, T. Vieira, E. de Barros Costa, and J. A. B. Moura

8.1 Introduction

Chatbots are increasingly cited to substitute human service agents in online shopping (Chen et al., 2023; Zogaj et al., 2023) becoming the first point of contact of the customer with the brand (Brady, 2018). Chatbots act as virtual agents that can support customers in their shopping experience by addressing their concerns and carrying out complex task autonomously (Pantano & Pizzi, 2020). Chatbots can offer personalised services in novel and entertaining ways (Chung et al., 2020). Owing to the volatile nature of this technology, pivotal research in the chatbot design calls for further research to optimise the use of this technology and highlights fashion-domain specific computational and non-computational research opportunities (Landim et al., 2022). AI-powered chatbots supported by large language models (LLMs), such as ChatGPT and Google Bard have gained attention by the creative industries in the last few years, disrupting services and transforming consumer behaviour, including what we know about Artificial Intelligence (AI).

V. Wanick (✉) · E. Bazaki (✉)
Winchester School of Art, University of Southampton, Southampton, UK
e-mail: vwv1n12@soton.ac.uk; E.Bazaki@soton.ac.uk

A. R. D. B. Landim · R. J. R. dos Santos · G. Paiva Medeiros · J. A. B. Moura
Graduate Program on Computer Science (PPGCC), Federal University of Campina Grande, Campina Grande, PB, Brazil
e-mail: andredantas@copin.ufcg.edu.br; roberiosantos@copin.ufcg.edu.br; gabriel.paiva@copin.ufcg.edu.br; antao@computacao.ufcg.edu.br

T. Vieira · E. de Barros Costa
Institute of Computing (IC), Federal University of Alagoas, Maceió, AL, Brazil
e-mail: thales@ic.ufal.br; evandro@ic.ufal.br

S. S. Muthu (ed.), *Sustainable Apparel Retailing*, Sustainable Textiles: Production, Processing, Manufacturing & Chemistry,
https://doi.org/10.1007/978-3-031-83875-0_8

Recently, studies have explored the value of emotional elements on the chatbot (Jiang et al., 2022; Yu & Zhao, 2024), and other human characteristics such as language (Boussakssou et al., 2022) chatbot anthropomorphization (Chen et al., 2023; Zogaj et al., 2023). Studies have also explored the chatbot in different contexts, chatbot advertising effectiveness (Van den Broeck et al., 2019), chatbot e-service, and customer satisfaction regarding luxury brands (Chung et al., 2020).

However, an AI-empowered sustainability chatbot with human elements to facilitate and enrich the customer experience is still under exploration. Human-like AI chatbots present new ways of enhanced customer-brand interaction and positive attitude towards receiving environmental and ethical information (Fuentes & Sörum, 2019). In fact, by having foundations based on psycholinguistic structures can humans and machines understand each other, thus providing a new paradigm to be considered in future research in the field (Kraljic & Lahav, 2024).

Although it may seem promising to add human effects to a chatbot, research has shown that when chatbots are 'too human' they might distract users from its real purpose (e.g., helping shoppers) (Brandtzaeg & Følstad, 2018); thus, balancing human and robot features is a key. Knowing the chatbot purpose also helps building specific relationships with consumers. For instance, chatbots designed with the aim to make recommendations can build intimate relationships with consumers, which leads to an emotional link (Silva & Canedo, 2024). Yet, challenges remain, particularly when adopting sustainable consumer behaviour values.

This chapter proposes an approach for the design of a sustainability-driven chatbot that could enable marketing managers to raise awareness, promote and understand sustainable consumption in the fashion industry, through value co-creative opportunities and human-centric elements. Based on (Landim et al., 2022) research on design approaches for the fashion industry chatbot and (Del Prete, 2022) work on sustainability chatbots, we propose a conceptual framework that encompasses these elements via suggested scenarios for applications. Future work is further discussed. In addition to this Introduction, the contents of this chapter are organised into three sections. The next section discusses chatbots in the fashion domain and their design aspects that lend them human-like features to make their conversations more easily acceptable and leading to more sustainable decisions by consumers. The discussion is centred on a virtues and values framework that promotes positive relationships between humans and chatbots. The following section explores how the implementation of chatbots may be applied to the fashion retail and supply chains to make the chains more sustainable. The last section offers conclusions and R&D gaps that may be fruitfully explored by considering voice input–output bots or based on the virtues and values framework to produce more appealing fashion chatbots'anthropomorphic designs.

8.2 Chatbots and Fashion Retailing

The fashion industry is being transformed by digital technologies and the growing popularity of e-commerce, social media, and metaverse platforms. Particularly during the COVID-19 pandemic, social distancing boosted sales through these platforms. This shift has created a need for effective communication with customers to meet their needs, preferences, and expectations. Fashion consumers are eager to adopt new technologies and digital platforms because they provide a more personalised and engaging online shopping experience.

Chatbots are often viewed as Conversational User Interfaces (CUIs). In fashion retail, they serve as a point of interaction at various stages of the customer journey (Lemon & Verhoef, 2016). The acceptance of chatbots has been analysed using models like the Technology Acceptance Model (TAM) and the Use & Gratification Theory (U&G), which primarily focus on perceived usefulness, ease of use, and enjoyment (Kasilingam, 2020; Rese et al., 2020). Beyond enjoyment and utilitarian aspects (Rese et al., 2020), the acceptance of chatbots also depends on the social presence often found in human-to-human interactions (Chaves & Gerosa, 2019). Factors such as privacy concerns and demographic differences also play a role in chatbot acceptance (Rese et al., 2020). Research on factors affecting the trustworthiness of chatbots includes aspects like anthropomorphism (Go & Sundar, 2019; Fatima et al., 2024; Chin et al., 2023), privacy (Cheng & Jiang, 2020), ethics (Shin et al., 2024), and even the elicitation of user emotions such as sadness and depression. There is a growing trend of using chatbots as value co-creators during the purchasing process, where they take on roles specific to fashion, such as a salesperson or stylist (Vaccaro et al., 2018). Rese et al. (2020) also noted gender differences in the use of chatbots among fashion e-commerce consumers, with a higher usage observed among women.

8.3 Towards a Human-Like Chatbot

To create chatbots that engage in more human-like conversations, whether through audio or text, remains a significant challenge. This progression relies heavily on artificial intelligence (AI) and machine learning (ML) algorithms. Although traditional ML techniques and rule-based systems are still prevalent in state-of-the-art chatbot research, modern chatbots require three essential components: a natural language understanding (NLU) unit to interpret the user's intent, a dialogue management (DM) system to generate a response based on the conversational context, and a natural language generation (NLG) component to craft responses in natural language (NL) (Suta et al., 2020). The perception of a chatbot's human-like qualities often depends on the user's interactions and the relationship formed with the chatbot. For instance, social chatbots can enhance affective responses and develop even friendships between humans and the AI (Brandtzaeg et al., 2022). A key component of a 'friendly AI' is reciprocity, but friendship is quite complex. (Fröding & Peterson, 2021) mention that in order to address such complexity, a 'friendly AI' should be rooted on virtues instead of values, which brings a lot of consideration towards the ethics of AI. However, the concept of 'friendly AI' has attracted criticism. A core problem is whether the humans interacting with the AI are still developing their moral tendencies and virtues an could be, therefore, impacted and influenced by the AI in an unethical manner (Li, 2021). Thus, having a mix of virtues and values as principles might be an alternative to CUI design. Central to CUI design, research in the area looked at the role of value-sensitive design (VSD) (Friedman et al., 2009) and Ethics (Lee et al., 2022; Wambsganss et al., 2021), giving consideration to many values, including credibility (De Souza Pereira Candello et al., 2023).

Human-like AI can be also investigated in other areas, such as videogames, via non-player characters (NPCs). In videogames, it was found that when NPCs move like players, they can be considered as 'human-like' agents (Milani et al., 2023). This suggests that there might be ways in which a chatbot can be felt to be more human due to its behaviour with technology. For instance, an indication that the chatbot is 'typing' some text may give novice users the perception that the chatbot is more like a human (Gnewuch et al., 2017). In fact, even making typing errors might make the chatbot more human-like (Bührke et al., 2021). Other insights from videogames might be applicable in this case. Aspects like personality traits, for instance, can be utilised to increase NPC assertiveness (Paradeda et al., 2019); facial features were also found to influence moral decision-making when players had to deal with in-game dilemmas (motivated by utilitarian or aggressive behaviours) (Ferstl et al., 2017).

Making a chatbot more 'human' means, therefore, that processes need to consider at least four aspects: (1) virtue and values modelling (via VSD principles), (2) personality traits, (3) relationship design (reciprocity, intimacy and similarity, empathy, self-disclosure, and trust) and (4) textual and non-textual cues (usually associated with the UI design). These elements, however, need to be adapted according to the consumer needs. For example, IKEA retired the chatbot 'Anna' in 2016 for 'being too human', as it felt disconnected from its purpose (deliver accurate and fast information to the consumer) (Brandtzaeg and Følstad, 2018, p.41).

8.4 Facilitating Sustainable Conversations: A Framework

Research on sustainability has become increasingly important in the fashion industry due to rising consumer concerns and regulatory pressure from governments and global organisations. Since recommendation systems can influence users' decisions, considering sustainability when making these are a promising approach to steer towards sustainable fashion consumption (Zhang et al., 2022). One of the problems is finding relevant sustainable fashion items in the marketplace (Skinner et al., 2023; Moon et al., 2015). This can be due to many factors such as consumer-facing accessibility of the e-retailing platform and filters to back-end product categorisation by the retailing team. Thus, discoverability of fashion brands and products, in this case depends not just on the platform itself but on a multichannel experience, which combines both offline and online touchpoints (Blázquez, 2014). However, one of the main issues is consumer sustainability knowledge, or in other words, what aspects that make a garment 'sustainable' as supply chain information, fabric, materials, recycling, and so on (Papasolomou et al., 2023). This means that beyond utilitarian and hedonic experiential needs, it is also important to give opportunities for consumers to learn more about the product. As Papasolomou et al. (2023)) mention that in order to shift towards sustainable consumption (particularly within the fast-fashion area), one needs to be aware that a product is indeed sustainable. It is also important to note that (Papasolomou et al., 2023) also found that there were gender differences regarding sustainability knowledge (e.g., women seemed to be more aware and would be willing to pay more for sustainable fashion brands).

However, how might this relate to virtues and values and the idea of building a relationship between humans and the chatbot? When looking at values, the framework 'LOV', developed by Kahle (1983) has been utilised to draw insights about sustainable consumer behaviour in a study with millennials consumers (Falke et al., 2022). Although, in fact, values influence sustainable consumption, most of the LOV values were self-centred and the study did not include the fashion industry. The values studied were sense of belonging, self-fulfilment, warm relationships, fun/enjoyment, security, self-respect, sense of accomplishment and excitement. Another study showed that both knowledge values and emotional values influence sustainable consumer behaviour (Sun et al., 2022), reinforcing that actual understanding of the impact of the product in the environment matters. But these values are still anthropocentric. This means that the relationship between humans and nature and the understanding of sustainability as a constrict could be re-defined. Thus, we suggest that other values-orientations, such as biocentric, values of solidarity and values of equality are added to 'anthropocentric' ones as suggested by (Torbjörnsson et al., 2011). For instance, biocentric values include reflections on nature protection and conservation, whereas anthropocentric values include exploitation of nature. This relates to Prete's approach to mindful consumption (Del Prete, 2022) for sustainability applied to the fast-fashion retailing industry and how chatbots can support this. In Prete's work, the chatbot functions as a recommender and there is less focus on relationships.

Considering these factors and reflections, there is a combination of factors that facilitate sustainable conversations between humans and the chatbot that needs to be considered and explored. This can be represented by four key areas (Fig. 8.1).

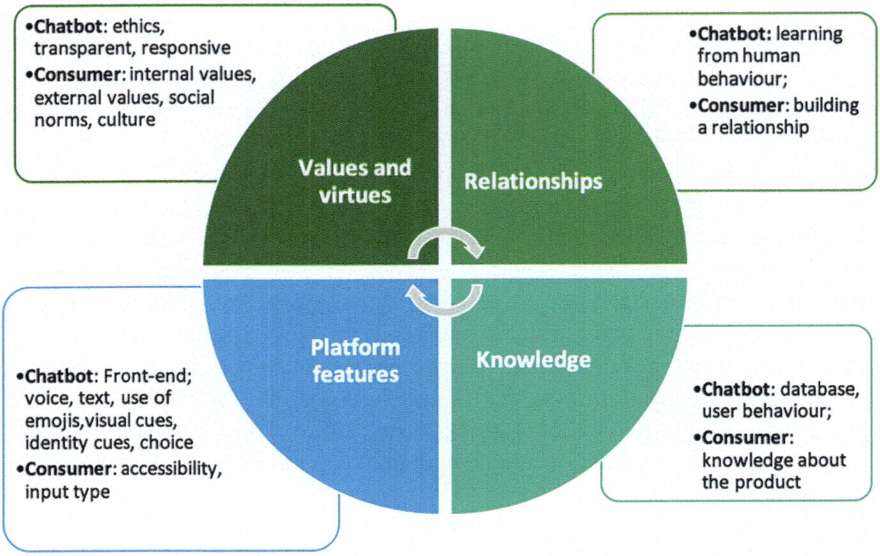

Fig. 8.1 The chatbot values and virtues framework

8.5 Values and Virtues

A core component of chatbot design is to be ethical and responsible. In this case, sustainable values need to be modelled with the chatbot features, respecting internal, external, and social values/norms. This also means that the chatbot needs to be transparent and ethical. For instance, if data is collected to inform their responses, there is a need to provide disclosure. But there is also a need to consider accessibility. A chatbot might recommend consumers 'recycle' their clothes, but it needs to include information on where and how this can happen, advantages and disadvantages of it, etc. The choice needs to remain with the consumer. Recent work on sustainability values highlights the integration of cultural values and their influence on sustainable consumption (Duong et al., 2024), anthropocentric values believing that nature should be preserved and protected owing to its utility to human (Sun et al., 2022), eco-centric (Bailey & Wilson, 2009) or intrinsic values (Justus et al., 2009) believing that nature should be protected due to its inherent value, biocentric values, values of solidarity and equality (Torbjörnsson et al., 2011). Kautish et al. (2024) examining the Biophilia hypothesis through the lens of values, beliefs, and norms theory found support for their hypotheses on the mediating role of nature connectedness and love for nature towards vegan good consumption and urge for nature-focused strategies. Therefore, it is possible to expect that the chatbot could lead the consumer into more meaningful conversations that strengthen consumer engagement and lead to more sustainable lifestyles.

8.6 Relationships

By framing the communication between consumers and chatbots via relationships, we move from transactional to relational experiences. Although the aim of the chatbot is not to become 'friends' with consumers, this aspect can be explored by CUI designers to make the experience enjoyable. Consumers are social beings and shopping with a companion is found to influence consumer experience positively (Merrilees & Miller, 2019). Adjusting social factors (e.g., including a friend in the conversation, or a group of friends) could affect user acceptance levels (Landim et al., 2022). Aspects like reciprocity, intimacy and similarity, empathy, self-disclosure and trust can be added in this context. In a customer–salesperson interaction, showing empathy, listening, being able to establish trust is particularly important in influencing decision-making (Chung et al., 2020). A study on luxury fashion brands e-service agents found that helpful, courteous, trustworthy interactions, entertainment, customisation, trendiness and problem-solving positively affect communication quality and customer satisfaction (Chung et al., 2020). This may be an opportunity for a sustainable *designbot* (Landim et al., 2022), functioning as a sustainable personal shopper with a particular personality trait based on consumer's preferences.

8.7 Knowledge

There are two ways to understand how knowledge can take place on facilitating sustainable conversations between consumers and chatbots. First, there is a need to categorise the database that will feed the chatbot according to what is considered sustainable or not. This can be due to materials, where the clothing was made, the supply chain information, etc. These aspects need to be inputted in the system so that the chatbot has enough knowledge to provide recommendations and information for consumers. The other side of the knowledge component is what the consumer knows about the brand or product and what they can learn whilst they interact with the chatbot. It might be possible that the chatbot can check what the consumer knows about the product, this can be via small polls, asking for feedback but also providing feedback for consumers. This component needs to work together with the other features from the CUI. At the same time, acquiring knowledge would need to satisfy the values and virtues from consumers. For example, learning a new idea or concept can fulfil not just utilitarian needs but also hedonic ones as learning can be fun and enjoying. For instance, the chatbot could reward the consumer by quizzing them. Here some gameful components can work to reinforce learning. However, this may influence the function of the chatbot and it should have a balance between how these values are portrayed, as some consumers are not looking for chatbots to 'make friends' with them but to provide them with informed choices as seen in Prete's work.

8.8 Platform Features

For the platform features, we take the categorisation proposed by (Landim et al., 2022) by adding knowledge and mindful consumption (Del Prete, 2022). Although 'understanding' is part of Landim's work, there was no understanding regarding the consumer or the non-computational factors. Values and virtues were also not included in Landim's work. Building on the work of Zogaj et al. (2023), chatbot's visual cues (i.e., anthropomorphization, gender) positively influence consumer behaviour. Anthropomorphic design elements increase consumer-chatbot self-congruence and lead to positive purchase intentions (Zogaj et al., 2023). Anthropomorphisation can be achieved by designing chatbots with human figures (i.e., visual cues) or giving them human names (i.e., identity cues) or adjusting conversational cues (i.e., language style) (Zogaj et al., 2023). Platform features such as emojis in hedonic chatbots are also found to enhance warmth and competence of the chatbot (Yu & Zhao, 2024). Gender is another highly highlighted platform feature (Landim et al., 2022; Papasolomou et al., 2023; Zogaj et al., 2023). Consumers are found to feel more similar with chatbots of their own gender (i.e., male consumers feel more comfortable interacting with male chatbots, similarly female consumers with female chatbots) (Zogaj et al., 2023). Yet, having a balance between hedonic and utilitarian values is crucial for chatbot effectiveness. For instance, Brandtzaeg and Følstad (2018) suggest that having 'less' components of a 'chat' and more towards a recommendation system (e.g., providing options/choices) might be a more efficient way to provide service-centric experiences. This may be possible due to the nature of the chatbot itself, which is different from what a salesperson would offer. Brandtzaeg and Følstad (2018) also highlight the use of images to provide recommendations to consumers (e.g., H&M chatbot allows users to send photos of garments to find similar ones).

8.9 Applications

A sustainable 'human-like' chatbot can be implemented in different layers of the consumer experience and across different channels that support the fashion retailing process, and in this case, circular fashion practises (Dissanayake & Weerasinghe, 2021). A circular fashion system has four sections: end-of-life, consumption, materials, and design, which are permeated by strategies and challenges, including logistics and supply chain management. Thus, in the next subsections we discuss the applications of our framework (Fig. 8.1) within circular fashion systems and potential applications (Fig. 8.2).

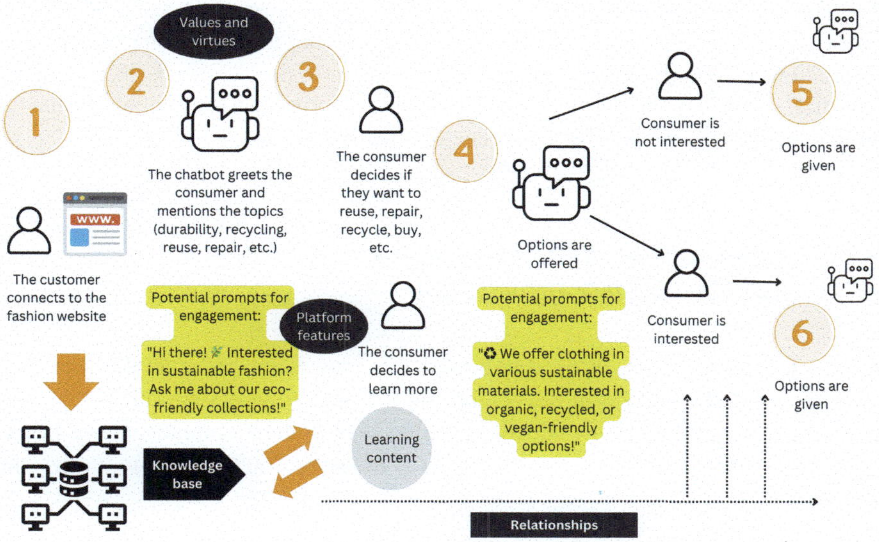

Fig. 8.2 Implementing the chatbot values and virtues framework based on the adapted approach from Del Prete (2022)'s mindful sustainable consumption through sustainable chatbots

8.9.1 Raising Awareness About Product Life Extension

Strategies that tackle end-of-life are usually driven by product life extension, such as reuse, remanufacturing, recycling, and reselling clothes (e.g., second-handed clothes). In this case, the knowledge base should be able to include aspects from the supply chain, in order to provide accurate recommendations for consumers. Suggestions and options can be given to consumers to educate them about end-of-life issues.

8.9.2 Providing Sustainable Choices

As noted in Brandtzaeg and Følstad (2018) and Del Prete (2022), giving a choice for consumers is crucial. Del Prete (2022) does that in their work, whilst presenting various choices for consumers before they decide to make a purchase. In our proposed approach (Fig. 8.2), the chatbot learns from consumers based on their interactions and give prompts in form of questions to allow consumers to learn more about a specific feature or perspective that relates to sustainability. This means that there is a learning opportunity in this scenario that can be addressed in the future. Equally, the chatbot can use a friendly language, including emojis. Options to repair, rent, swap or lease clothing can be given at this stage. At the same time, it is possible to determine consumer preferences and behaviour through machine learning (ML) mechanisms. These can be added to the chatbot development, but it should also be transparent. The consumer should be aware that in order to get personalised recommendations they might need to give consent to utilise their data. If possible to create a profile, consumers could visualise their carbon footprint, for example, or maybe get an information about how they relate to other consumers.

8.9.3 *Filtering Through Material, Carbon Footprint and Sustainable Information*

Information about materials need to be added to knowledge base and the choice of materials should be added in the chatbot. In this scenario, an opportunity to learn more about a material can be added or to filter results and recommendations based on a specific material. This means that fashion items need to be added to the system and tagged accurately.

8.9.4 *Revising the Supply Chain and Fashion Design Practises*

Interactions of consumers with the chatbot might give insights for fashion brands about their choices and whether these can improve fashion design. Trends can be added and scrapped across the web to communicate with consumers. For that, data from consumer preferences could feed into the chatbot knowledge base. Depending on consumers' choices and what is available in the knowledge base, this may influence fashion design in a larger scale, including the supply chain. In this case, we envision that the consumer interaction with the chatbot may influence design for recycling, composting, disassembly, customisation, and longevity.

8.10 Conclusion and Future Work

Considering this, we expect our framework can demonstrate how these tools can encourage the adoption of sustainable practises, inspire consumers to make eco-friendly choices, and thus advance environmental sustainability. This chapter aims to provide practical guidelines for leveraging AI in promoting sustainability within the fashion industry. These guidelines are intended to be useful for academics, industry professionals, and policymakers. We hope that this research will help stakeholders better understand how AI can transform sustainability in fashion, leading to a more sustainable future for the industry and the circular economy. Managers are recommended to start with selecting the type of chatbot, hedonic vs. functional, to research and investigate the values and virtues appreciated by the customers, the type of relationships, functions, and knowledge base required. We also suggest that managers look into the learning aspects of sustainable consumer behaviour that could be enhanced via the chatbot design and implementation. For example, consumers might not be aware that their choice of product might have an impact on the whole process, from supply chain to product end-of-life, but equally they might not be aware that a specific garment is made with sustainable resources, or in a specific country; consumers might not know that the company offers a space to recycle elements of their garments or a clothing swapping service. Thus, the chatbot can make these external services more evident, and educate the consumer towards a more sustainable choice. This not only pushes the consumer to make an informed decision, but it also motivates companies to provide more sustainable services and align their ethos towards the circular economy (and perhaps let them revise their business model). In the end of the day, it is the consumer who makes the choice, not the chatbot. In the future work, we recommend focusing on the interplay between self-concept, values, and virtues as a starting point for the anthropomorphic design cues. It is also important to mention that we have addressed the concepts of chatbot via a text perspective. Thus, further work can look at the influence of voice-interface design in consumer behaviour, particularly when utilising different 'human' voice tones and accents.

References

Bailey, I., & Wilson, G. A. (2009). Theorising transitional pathways in response to climate change: Technocentrism, ecocentrism, and the carbon economy. *Environment and Planning A, 41*(10), 2324–2341. https://doi.org/10.1068/a40342

Blázquez, M. (2014). Fashion shopping in multichannel retail: The role of technology in enhancing the customer experience. *International Journal of Electronic Commerce, 18*(4). https://doi.org/10.2753/JEC1086-4415180404

Boussakssou, M., Ezzikouri, H., & Erritali, M. (2022). Chatbot in Arabic language using seq to seq model. *Multimedia Tools and Applications, 81*(2), 2859–2871. https://doi.org/10.1007/s11042-021-11709-y

Brady, M. K. (2018). How can looking at the whole picture help you serve customers? In *Mapping out marketing: navigation lessons*. Routledge. https://doi.org/10.4324/9781315112602-28

Brandtzaeg, P. B., & Følstad, A. (2018). Chatbots: Changing user needs and motivations. *Interactions, 25*(5), 38–43.

Brandtzaeg, P. B., Skjuve, M., & Følstad, A. (2022). My AI friend: How users of a social chatbot understand their human-AI friendship. *Human Communication Research, 48*(3), 404–429. https://doi.org/10.1093/hcr/hqac008

Bührke, J., et al. (2021). Is making mistakes human? On the perception of typing errors in chatbot communication. In *Proceedings of the annual Hawaii international conference on system sciences*. https://doi.org/10.24251/hicss.2021.541

Chaves, A.P., & Gerosa, M.A. (2019). How Should My Chatbot Interact? A Survey on Social Characteristics in Human–Chatbot Interaction Design. *International Journal of Human–Computer Interaction, 37*, 729–758.

Chen, J., Guo, F., Ren, Z., Wang, X., & Ham, J. (2023). Human-chatbot interaction studies through the lens of bibliometric analysis. In *Universal access in the information society*. Springer. https://doi.org/10.1007/s10209-023-01058-y

Cheng, Y., & Jiang, H. (2020). How Do AI-driven Chatbots Impact User Experience? Examining Gratifications, Perceived Privacy Risk, Satisfaction, Loyalty, and Continued Use. *Journal of Broadcasting & Electronic Media, 64*, 592–614.

Chin, H., Song, H., Baek, G., Shin, M., Jung, C., Cha, M., Choi, J., & Cha, C. (2023). The Potential of Chatbots for Emotional Support and Promoting Mental Well-Being in Different Cultures: Mixed Methods Study. *Journal of Medical Internet Research, 25*.

Chung, M., Ko, E., Joung, H., & Kim, S. J. (2020). Chatbot e-service and customer satisfaction regarding luxury brands. *Journal of Business Research, 117*, 587–595. https://doi.org/10.1016/j.jbusres.2018.10.004

De Souza Pereira Candello, H. C., et al. (2023). "This means nothing to me": Building credibility in conversational systems. In *Proceedings of the 5th international conference on Conversational User Interfaces, CUI 2023*. https://doi.org/10.1145/3571884.3603759

Del Prete, M. (2022). Mindful sustainable consumption and sustainability chatbots in fast fashion retailing during and after the COVID-19 pandemic. *Journal of Management and Sustainability, 12*(1), 19. https://doi.org/10.5539/jms.v12n1p19

Dissanayake, D. G. K., & Weerasinghe, D. (2021). Towards circular economy in fashion: Review of strategies, barriers and enablers. *Circular Economy and Sustainability, 2*, 25–45.

Duong, C. D., Le, T. L., Lee, E.-M., & Gadomska-Lila, K. (2024). How cultural values integrate with each other to trigger sustainable consumption: A cross-culture study. *Asia Pacific Journal of Marketing and Logistics, 36*(4), 814–836. https://doi.org/10.1108/APJML-05-2023-0484

Falke, A., Schröder, N., & Hofmann, C. (2022). The influence of values in sustainable consumption among millennials. *Journal of Business Economics, 92*(6), 899–928. https://doi.org/10.1007/s11573-021-01072-7

Ferstl, Y., Kokkinara, E., & McDonnell, R. (2017). Facial features of non-player creatures can influence moral decisions in video games. *ACM Transactions on Applied Perception, 15*(1), 1–12. https://doi.org/10.1145/3129561

Friedman, B., Kahn, P. H., & Borning, A. (2009). Value sensitive design and information systems. In *The handbook of information and computer ethics*. Wiley. https://doi.org/10.1002/9780470281819.ch4

Fröding, B., & Peterson, M. (2021). Friendly AI. *Ethics and Information Technology, 23*(3), 207–214. https://doi.org/10.1007/s10676-020-09556-w

Fuentes, C., & Sörum, N. (2019). Agencing ethical consumers: Smartphone apps and the socio-material reconfiguration of everyday life. *Consumption Markets & Culture, 22*(2), 131–156. https://doi.org/10.1080/10253866.2018.1456428

Go, E., & Sundar, S.S. (2019). Humanizing chatbots: The effects of visual, identity and conversational cues on humanness perceptions. *Comput. Hum. Behav., 97*, 304–316.

Gnewuch, U., et al. (2017). "The chatbot is typing ..."—The role of typing indicators in human-chatbot interaction. In *Proceedings of the sixteenth annual pre-ICIS workshop on HCI research in MIS*.

Jiang, C., Zhang, C., Ji, Y., Hu, Z., Zhan, Z., & Yang, G. (2022). An affective chatbot with controlled specific emotion expression. *Science China Information Sciences, 65*(10). https://doi.org/10.1007/s11432-020-3356-4

Justus, J., Colyvan, M., Regan, H., & Maguire, L. (2009). Buying into conservation: Intrinsic versus instrumental value. *Trends in Ecology & Evolution, 24*(4), 187–191. https://doi.org/10.1016/j.tree.2008.11.011

Kahle, L.R. (1983) Social Value and Social Changes. New York: Praeger.

Kasilingam, D. (2020). Understanding the attitude and intention to use smartphone chatbots for shopping. *Technology in Society, 62*, 101280.

Kautish, P., Thaichon, P., & Soni, P. (2024). Environmental values and sustainability: Mediating role of nature connectedness, and love for nature towards vegan food consumption. *Journal of Consumer Behaviour, 23*(3), 1130–1145. https://doi.org/10.1002/cb.2267

Kayeser Fatima, J., Khan, M.I., Bahmannia, S., Chatrath, S.K., Dale, N.F., & Johns, R. (2024). Rapport with a chatbot? The underlying role of anthropomorphism in socio-cognitive perceptions of rapport and e-word of mouth. *Journal of Retailing and Consumer Services.*

Kraljic, T., & Lahav, M. (2024). From prompt engineering to collaborating: A human-centered approach to AI interfaces. *Interactions, 31*(3), 30–35.

Landim, A. R. D. B., et al. (2022). Chatbot design approaches for fashion E-commerce: An interdisciplinary review. *International Journal of Fashion Design, Technology and Education, 15*(2), 200–210. https://doi.org/10.1080/17543266.2021.1990417

Lee, M., et al. (2022). Ethics of conversational user interfaces. In *CHI conference on human factors in computing systems extended abstracts* (pp. 1–7). ACM. https://doi.org/10.1145/3491101.3503699

Lemon, K. N., & Verhoef, P. C. (2016). Understanding customer experience throughout the customer journey. *Journal of Marketing, 80*(6). https://doi.org/10.1509/jm.15.0420

Li, O. (2021). Problems with "friendly AI". *Ethics and Information Technology, 23*(3), 543–550. https://doi.org/10.1007/s10676-021-09595-x

Merrilees, B., & Miller, D. (2019). Companion shopping: The influence on mall brand experiences. *Marketing Intelligence & Planning, 37*(4), 465–478.

Milani, S., et al. (2023). Navigates like me: Understanding how people evaluate human-like AI in video games. In Conference on human factors in computing systems—Proceedings. https://doi.org/10.1145/3544548.3581348

Moon, K. K. L., Lai, C. S. Y., Lam, E. Y. N., & Chang, J. M. (2015). Popularization of sustainable fashion: barriers and solutions. *The Journal of the Textile Institute, 106*(9), 939–952.

Pantano, E., & Pizzi, G. (2020). Forecasting artificial intelligence on online customer assistance: Evidence from chatbot patents analysis. *Journal of Retailing and Consumer Services, 55*, 102096.

Papasolomou, I., Melanthiou, Y., & Tsamouridis, A. (2023). The fast fashion vs environment debate: Consumers' level of awareness, feelings, and behaviour towards sustainability within the fast-fashion sector. *Journal of Marketing Communications, 29*(2), 191–209. https://doi.org/10.1080/13527266.2022.2154059

Paradeda, R., et al. (2019). The role of assertiveness in a storytelling game with persuasive robotic non-player characters. In *CHI PLAY 2019—Proceedings of the annual symposium on computer-human interaction in play*. https://doi.org/10.1145/3311350.3347162

Rese, A., Ganster, L., & Baier, D. (2020). Chatbots in retailers' customer communication: How to measure their acceptance? *Journal of Retailing and Consumer Services, 56*, 102176.

Shin, D., Lee, J.H., & Noh, W.I. (2024). Realizing Corrective Feedback in Task-Based Chatbots Engineered for Second Language Learning. *RELC Journal.*

Silva, G. R. S., & Canedo, E. D. (2024). Towards user-centric guidelines for chatbot conversational design. *International Journal of Human–Computer Interaction, 40*(2), 98–120. https://doi.org/10.1080/10447318.2022.2118244

Skinner, E. A., Le Busque, B., Dorrian, J., & Litchfield, C. A. (2023). # sustainablefashion on Instagram: A content and network analysis of user-generated posts. Journal of Consumer Behaviour, 22(5), 1096–1111.

Sun, J., et al. (2022). Substantiating nexus between consumption values and sustainable consumption behavior: A way toward sustainable business. *Frontiers in Psychology, 13*, 908391. https://doi.org/10.3389/fpsyg.2022.908391

Suta, P., Lan, X., Wu, B., Mongkolnam, P., & Chan, J.H. (2020). An Overview of Machine Learning in Chatbots.

Torbjörnsson, T., Molin, L., & Karlberg, M. (2011). Measuring attitudes towards three values that underlie sustainable development. *Utbildning & Demokrati – tidskrift för didaktik och utbildningspolitk, 20*(1), 97–121. https://doi.org/10.48059/uod.v20i1.945

van den Broeck, E., Zarouali, B., & Poels, K. (2019). Chatbot advertising effectiveness: When does the message get through? *Computers in Human Behavior, 98*, 150–157. https://doi.org/10.1016/j.chb.2019.04.009

Vaccaro, K., Agarwalla, T., Shivakumar, S., & Kumar, R. (2018). Designing the Future of Personal Fashion. Proceedings of the 2018 CHI Conference on Human Factors in Computing Systems.

Wambsganss, T., et al. (2021). Ethical design of conversational agents: Towards principles for a value-sensitive design. In *Innovation through information systems. WI 2021. Lecture notes in information systems and organisation*. Springer. https://doi.org/10.1007/978-3-030-86790-4_37

Yu, S., & Zhao, L. (2024). Emojifying chatbot interactions: An exploration of emoji utilization in human-chatbot communications. *Telematics and Informatics, 86*, 102071. https://doi.org/10.1016/j.tele.2023.102071

Zhang, T., Feng, C., Chen, H., & Xian, J. (2022). Calming the customers by AI: Investigating the role of chatbot acting-cute strategies in soothing negative customer emotions. *Electronic Markets, 32*, 2277–2292.

Zogaj, A., Mähner, P. M., Yang, L., & Tscheulin, D. K. (2023). It's a match! The effects of chatbot anthropomorphization and chatbot gender on consumer behavior. *Journal of Business Research, 155*, 113412. https://doi.org/10.1016/j.jbusres.2022.113412

Index

S. S. Muthu (ed.), *Sustainable Apparel Retailing*, Sustainable Textiles: Production, Processing, Manufacturing & Chemistry,
https://doi.org/10.1007/978-3-031-83875-0

The manufacturer's authorised representative in the EU is Springer Nature Customer Service Centre GmbH, Europaplatz 3, 69115 Heidelberg, Germany. If you have any concerns regarding our products, please contact ProductSafety@springernature.com

Printed and bound by CPI Group (UK) Ltd, Croydon, CR0 4YY
01/10/2026
02225980-0010